Building the Future

Building the Future

Mac and Sam

VANTAGE PRESS
New York

FIRST EDITION

Published by Vantage Press, Inc.
419 Park Ave. South, New York, NY 10016

Manufactured in the United States of America
ISBN: 978-0-533-15680-1

Library of Congress Catalog Card No.: 2006909887

0 9 8 7 6 5 4 3 2 1

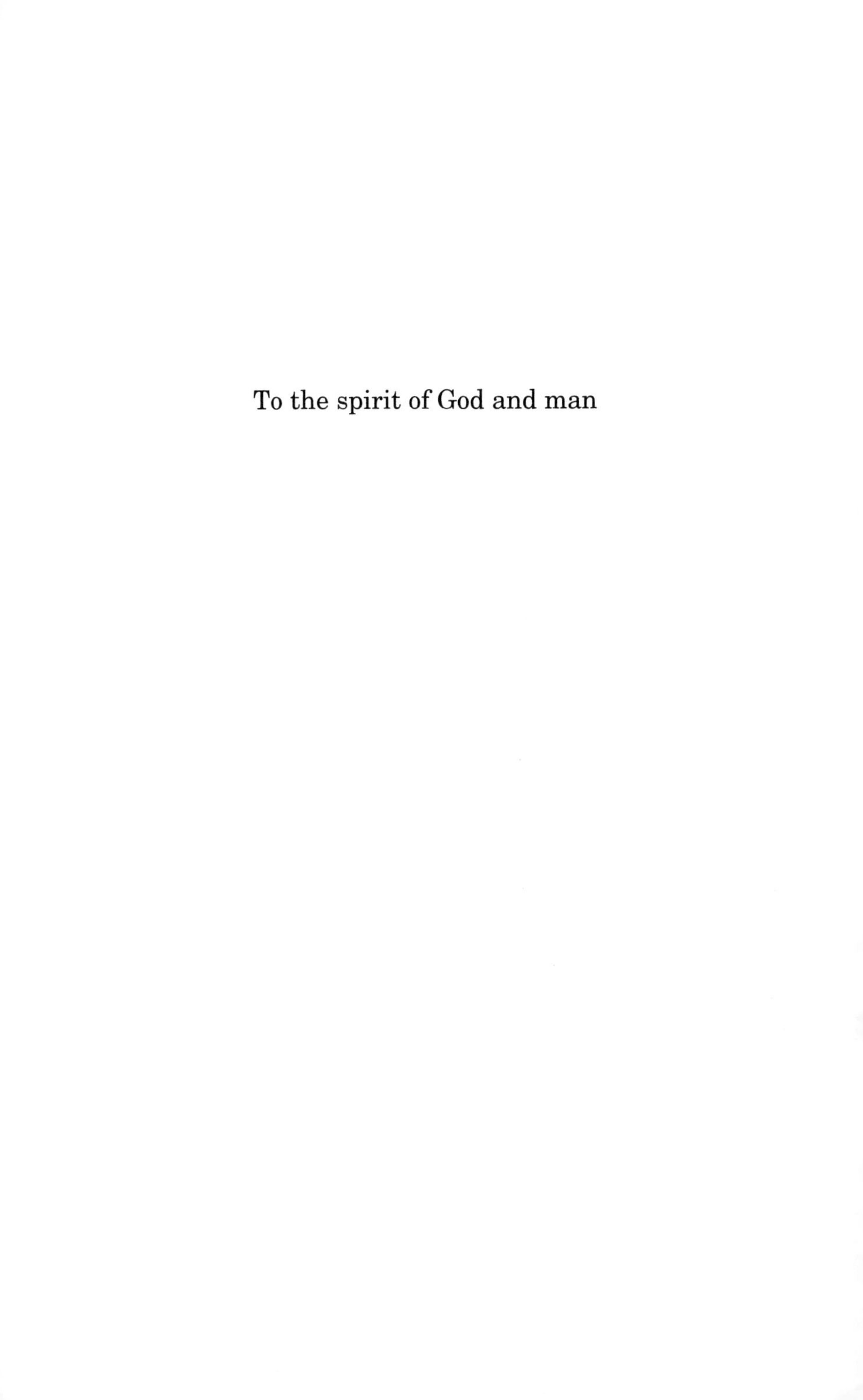

To the spirit of God and man

Contents

Credits and Acknowledgments

This book is a little different than most. As far as the writer knows, it could never have been prepared without all of the past and the hopes for the future. I suppose that means the credit goes to God; all of man's past; and most of man's hopes for the right decisions going into and through the future.

Introduction

The future . . . if the next 500 years are anything like the past 500 years, WOW! Fantastic and amazing things/miracles are sure to happen. Imagine for a few moments the rates of change, growth and progress in the United States from the time the Pilgrims landed—back in the 1600s. Take some time, use your imagination and apply those rates of change to the next 500 years. One point of view wants to tell us that, here in 2007, we are just laying the groundwork for a portion of the future and that we have only scratched the surface of things to come.

Each year new inventions and discoveries develop and come to market. When they do, it is for mankind's benefit; helping to raise the standard of living, improving the quality of life, and contribute to economic growth and stability. The great thing is that there is always a new invention or major discovery just around the corner!

In this book we are going to explore interesting ideas about man's past, present, and future. The primary focus is on the development of new methods of expansion and construction that will help insure the economic generators of the world in the future. Discussing methods that, at this time, are part fact and part science fiction.

The writing is pretty straightforward and should be easy to read and understand. Hopefully, the book will provide some positive ideas and inspire more of those ideas in adults and children. In a time when both children and adults are bombarded with so much negative, confusing, and destructive input, the author wanted to provide something constructive. The composition of the writing is, to say the least, different—both in thought and structure. There are few wasted words and I have tried to keep the rambling, wild tangents and trips into far left field to a bare minimum.

At times throughout the book I will ask the reader to think, consider, and ponder. The reason is simply because most readers will have volumes of knowledge in memory to draw from; much more than I can put into words. Keep in mind the focus points: new expansion methods, economic stability and economic growth. The old sayings, comparisons, examples, and personal thoughts help present the stories.

It's not every day that someone wants to tell you (and sell you) a story about building cities above the coastal seas and manufactured real estate. Miracles happen every day and more than likely, the next one is just around the corner. As an example, take a moment and imagine being able to tell the Pilgrims how things would turn out here in 2007. Miraculous, might be a reply; or maybe, you're pulling my leg could be another. Many miracles around a lot of corners in the past few hundred years. Look toward the future and try to imagine one of your offspring being able to come back and tell you of things that await the kids in those days! Let's hope that it is all good.

Back when the United States was getting started, one of the census reports shows a population of about 3.9 million people in 1790 and six cities of 8,000 or more. In 2000 (that is only 210 years later) the census report shows the population of the U.S. at about 284 million and the largest city area (New York) is shown as having about eight million people. That is a very large increase. The

One to Grow On
(Story Number One)

One
The What

This story is based on a theory that involves the construction of massive concrete and steel structures that rise from the coastal areas' sea floor, to build and expand the cities of the future on. The theory is to design the processes and methods for such construction; it might be described as manufactured real estate that is mass-produced, with designs that include using all of the combined experience and knowledge of past and present construction in and around the sea. Construction experience that includes dams, bridges, shipping harbors and docks, under-sea tunnels, river locks, offshore drilling platforms, continuous mass production standards; along with anything else that will help man. Including quite a few prayers for God's help in the matter.

One of the most exciting parts of this idea of building real estate above the sea is, that almost every major coastal city in the world could benefit from it at some point in time. History has taught us that real estate, as a rule, continually increases in value. These increases help to generate wealth—financial, economic growth and stability, along with all of the benefits associated with these.

Common sense tells us, and most economic experts would agree, that if man can successfully engineer a profitable method of construction, in and above the coastal seas of the world, it would probably be one of the profitable engineering miracles ever produced! We can gain some understanding of manufactured real estate when we consider high-rise buildings and skyscrapers. These types of structures have developed over the past 100 years or so (remember, in the time of man, one hundred years does not really amount to much). Some of these structures now reach heights of one-quarter to one-half mile, straight up! The value generated by this type of construction is a study all of its own. We should be able to easily agree that tremendous economic supply and wealth have been generated with this type of construction. When we start combining the engineering expertise that raises structures to such heights with technological advances that have created the success of the space programs, plus long-term economic supply and demand statistics combined with population growth—the idea of building structures in the coastal sea areas becomes more realistic.

If we look back through time, someplace in there, we can find that man has done almost anything we can imagine with this piece of real estate that we call the United States. Going way back, it was just exploration, hunting, and gathering food. Then he learned to grow crops; soon after that he began harvesting the timber and clearing the forests. He has moved and leveled mountains. He had drilled it, drained it, and flooded it. His largest city area now has close to seventeen million people. He has built and constructed things almost beyond belief—all in just a few hundred years! Some of it is pretty amazing stuff!

Science tells us the surface of the earth is about thirty percent dry land, with about seventy percent covered with water. In the chronological order of development, from hunting and gathering, to farming, construction, population, and overcrowded cities—it almost makes perfect sense that one of man's next jobs will involve large-scale, heavy and long-lasting, construction above some of the water. The water that covers seventy percent of the surface of the earth.

Taking a viewpoint of supply and protection, along with the anticipated economic needs of the future, when populations of billions begin to double, this method of expansion and construction starts to feel as though it will be demanded by necessity, and not by choice.

Manufacturing real estate that is almost hurricane-proof, by the square mile, in prime market coastal areas! I would like to think that one of the best parts of this type of expansion and construction is that it holds the potential for raising the standard of living and quality of life for millions and eventually billions of people worldwide.

All through time and history, far-fetched ideas have been dreamed up, proposed, and developed. With many of these types of proposals, conventional thought and wisdom almost immediately want to tell us No! It is impossible; it would cost too much; it is folly or speculative science fiction, etc., etc. Then, with a little time, study and investment, industry, business and governments pound out solutions and make things work. One of the very best examples to illustrate this thought, from recent history, is the space program.

Imagine the possibility of creating another brand new industry. A new industry that could have the worldwide impact, importance, and economic supply and demand effects that might compare with the automotive and transportation industries. Then add that quantity of economic power into the economy of the U.S. and the rest of the world. As history, time and experience have proven through the years, we can only imagine or guess at how it might develop, over the next one hundred or two hundred years. A serious question for some of the readers is this: could this process develop into one of, if not "the," critical economic suppliers that the future will require just to survive?

Let us suppose for a while. Suppose the idea gets the necessary support; a few engineers and accountants start looking into it, and it starts to gain some momentum. That is all it would take. History has proven again and again, that is how things get started. Within years (of recent history) billions of dollars of investment and funding make big things happen, fast! With construction of this type, quantities are huge; machinery is big, heavy and expensive. Let's make a comparison. The automobile industry started about one hundred years ago. It did not exist. Look at it today—it has continuously increased in size to this very day. It has provided the load on primary industry that has built the economic systems of the world. Imagine the type of construction that we are discussing; repeating and doubling in size in a comparable manner—one word, Big!

Good things can happen. This thing could hold the potential to cure a lot of problems, as long as it is developed and managed properly. Imagine a whole new financial and investment sector, good high paying career and employment opportunities, tax revenue increases, retirement plans—just to name a few. A new industry not only for the U.S., but one that can spread around the world. The positive effects could apply to almost every industry and business. Again, consider and think of the far-reaching effects of the automotive and transportation industries.

Serious large scale use of the type of construction

that we are discussing holds the potential to create the financial boom of all booms. The kind that can sustain, increase and repeat for hundreds, then thousands of years. By stressing monetary benefits at the beginning of the story helps to attract interest; almost everyone could use a little more money. But the concept is so much more than making a buck. The effect of large scale use could have far-reaching influence that move into areas that include national security, independence, freedom, and survival of humanity as we know it today.

In a time when Uncle Sam is having trouble giving money away at historically low interest rates, business, industry and government should jump on this thing with both feet. To, at least, give it a chance for development. In the world we have some very powerful, influential and wealthy people. People (and companies) that have the ability to get the ball rolling on projects such as this with a telephone call. All it takes is a little information, and a little prodding, to supply some seed money. When we consider throwing seed money at business and industry, about the only thing that is required for success is demand for the product. Any product, even manufactured real estate or manufactured islands in the coastal seas.

Creating demand. The great thing about land, property, or real estate, is the demand for it is already created. It exists and it started a long, long time ago. History provides all of the actual facts of how the demand for real estate has continued, grown or increased all through the years. The same historical facts also prove, beyond any doubt whatsoever, that the value increases with the demand. Now, here in 2007, prime real estate in the U.S. is some of the most valuable on earth. We can be pretty sure that, as history has proven, the value will continue to increase from here on out. We can almost say that, because of these facts, business, industry and government are required or have a duty to study a proposed method of manufacturing it above the surface of the coastal seas—at some point in time. Primarily, because of the economic supply that can be created—the supply that feeds the world.

The demand to manufacture in this manner can be compared to the demands to manufacture with all of our primary industries and processes. In the early twentieth century man didn't have many automobiles. The demand for them was just getting started. The automobile did develop, and evolved into the annual multi-billion dollar industry that it is today; the process took about one hundred years. In 1850 man didn't need, and a large percentage didn't want, an automobile (how that has changed). Here in 2007, the automobile industry is a critical portion of all human life on earth. The automobile that started out as a novel or luxury machine, now demands that man build and use it just for survival. Man does not have a choice; it must be manufactured! This required demand developed in less than one hundred years! The real estate industry developed in a similar manner.

Back in 1850 old Uncle Sam gave man land for free. It didn't take long for that to change! Soon, the demand for real estate began to feed on itself. Value was created and continued to increase. Here in 2007, real estate, like the automobile in the preceding example, demands that man do something with it in order to survive. He must; he has no choice in the matter. Man is required to continuously process real estate. He must farm it; he must mine it; develop and recover the resources; build on it; sell it; buy it; trade it; drain it; flood it; and build highway systems on, above and beneath it. All of these things are demanded to happen in order for the economic systems to function.

The sea is just as sneaky. The sea demands that man use it in order to survive and make his economic systems function. It was all pretty simple in the beginning; man started out catching fish for food and bathing in the sea. Then it was rafts, sponges, and boats; more boats and bigger boats; then more bigger boats and more and bigger fish. He must, with no choice in the matter, use and develop the natural resources that it provides in order to supply the needs of mankind. The idea that we are discussing, construction in and above the sea, is beginning to take on some of these same must-do characteristics. Kind of like a new demand that starts in the twenty-first century.

All through the years man has created, invented and developed stuff, and some pretty amazing stuff. Most of the time it was to meet or supply a demand. But, with a little bit of study, we see that almost each creation evolves into another new or different demand that requires further invention, creation, development and production. It appears to be one of the oldest constants or natural processes, not only of things, but of human development. History supplies the proof that demonstrates once man makes a working model of a new design, invention or process—that it is only a matter of time until it is required and demanded to be produced to meet the needs of mankind. These things indicate and seem to want to tell us a few things. Such as the fact that we have this idea in print, almost. It says to us that it is in man's future—that when he raises his first modern city on a structure that rises from the sea floor, it will only be a matter of time until more and more of those structures are demanded worldwide.

Is the demand more real and closer than we might

think? Some of the successful economic and monetary balancing acts by governments and industry during the past few years have been close to the miracle classification. In that they prevented the deep economic depressions that could have destroyed the world's economic systems. There were many times when it seemed as though one false move would have sent all of the stock markets straight to the bottom. That doesn't mean it can't happen again. It does mean that mankind must constantly search for and develop new ways and methods to deal with the economic supply and demand problems. They will never decrease and will forever increase at larger and faster rates.

If or when man is ready to sell a project such as this to investors, industry and government, the indicators tell us that it will be much easier when the markets are up and stable. From one point of view (in the spring of 2005) investors, the markets and the economic systems were searching and hoping for something that could carry this much weight. When the business cycles are low or at the bottom and money is tight, new projects can be hard to sell. That would be considered as too late.

We have covered a lot of ground in these first few pages, touching on a variety of thoughts and subjects to help define what this, One to Grow On, is all about. We can select a portion of the bottom line that says it is about doing good and the right thing. But we better not leave out the fact that Mac and Sam (and the writer) are attempting to make a little money by getting the book published.

This book is dedicated to the spirit of God and man—we haven't discussed that yet. Now might be a good time. The spirit is powerful stuff. It is one of those things that science will always have trouble with because it can't be defined with numbers or a formula. People of faith don't have much trouble understanding the spirit; and God and the supernatural power are not usually shown in the material lists and design specifications of engineering plans—that is fairly easy to understand. But, that does not mean that the spirit is not involved. So in a work such as this, we are permitted to introduce the unknown supernatural power of God into the design specifications. From the beginning of man's written history he has prayed or called out to God and the spirit, especially in times of need or want. Leaders, rulers, kings, presidents, preachers, mothers, fathers and children have asked the unknown supernatural power of God for help, strength, power, and guidance to manage the needs of life and survival. Sometimes it works.

Some of the best human minds known to man have held strong spiritual beliefs. We should learn from this. It indicates, or tries to teach and reinforce the teachings that tell us, if God is with us, who can be against us? With all of this in mind as we think, wonder, ponder the present, the future and the past—we introduce an idea for the future of man that has him constructing and raising cities on structures that rise from the coastal sea floors. This in turn can open the door to seemingly endless thoughts and questions. Before we get all tangled up within these questions and thoughts, sometimes it is good to grab onto some faith, hook up with the spirit, and remember the teaching that says, if God is with us, who can be against us?

The ability to design and construct foundation structures that can support the weight/load of a modern city and not be destroyed by the forces of the coastal sea locations is not a small or easy task Expert, precision engineering of the quality that puts men on the moon; remote controlled machinery used on Mars, that splits atoms, and raises skyscrapers one-half mile straight up is needed.

Think forward into the future. Let's pick the year 2300. History wants to tell us that there might be a couple of billion people in the U.S. by then; with that, the imagination can run wild. Between now and then, financial and economic demands will grow and continuously demand more and more and more. One of the questions man will need to ponder is, could the science, methods and abilities to construct these types of foundation structures actually carry the financial and economic weight of the world in those days, in addition to one small developmental test projects that could be started tomorrow?

If you are anything like the writer—part of us is whispering, 2300, ha—we will be long gone by then. But, just maybe, man has a chance at getting this thing started in the near future. Suppose that some of the big money managers and decision makers start making a few arrangements to take a shot at this. Expert representatives of engineering, finance, industry, government, and labor all get on board and plan starts to come together—this thing is going to happen. Finance and government dipped into the piggy banks and found a few hundred billion to throw into it; all of the other sectors are looking for something to do anyway, so they are all ready.

At this point it is a matter of getting engineering to make the proper designs, tests and final start-up plans. Someplace in the design process a demonstration project, as history teaches us, is finally developed. It is a big deal; a very, very, big deal. This demonstration project holds the power to open a new era for the world that compares, in effect and on some scale, to things such as the Industrial Revolution, rocket science, electrical power in industry, the automobile, the transistor, etc. It is important. It would be a place to grow. It would be the proverbial, "One To Grow On." It could very well be the doorway to prosperity; and the possibilities that are available, the opportunities that could be created? This one has a shot at helping to end wars and making man live in peace. That means the demonstration project, the show piece, is a model of a modern city of the future—designed for millions of visitors from all corners of the world, to see it, as well as for business leaders, government leaders, educators and kids to visit.

Two of the most important needs for such a demonstration project would be good yearlong weather and construction that is almost hurricane-proof; it could withstand the worst storms with minimal damage. A favorite thought of the writer is to locate such a demonstration project on the east coast of Florida within viewing distance of nighttime space shuttle launches. This location represents a place on earth where all of the best of man's past, present, and future meet. It is a place where some of the early explorers arrived in their leaky wooden sailboats; it is a place where modern man leaves the earth to explore space; and if the demonstration project is built, it will be a place where man returns to the sea to build part of the future. Plus the weather is pretty good in that area, although farther south is usually warmer in the wintertime.

So, just how big do we build this thing? Well, like the old-timers used to say, that is the fifty dollar question, son, and a good one. Experienced professional engineers design success. They are required to determine the design. But, we can speculate, daydream and imagine. We can pick an arbitrary size of one hundred square miles as a size of use in our story; something to talk about. Say, ten miles wide and ten miles long. That is a lot of area and could provide for a considerable amount of construction. If the real thing ever happens it might be smaller or much larger; we will just have to wait and see.

One hundred square miles is a deceptive size. If we think of 100 square miles in open prairie and ranch land, it doesn't seem like much. If it was ten miles square and had a road across it, we could drive across it in about ten minutes. In the open country it's a drop in the bucket. But, when compared to some of the major cities, it takes on a whole different meaning. At those locations it can be the home of millions of people.

As an example: the largest city in the U.S., New York, is about 300 square miles with a population of eight million people. Washington, D.C. is about 68 square miles. If we think of a mass of concrete and steel that forms a city ten miles wide and ten miles long, located in the vast area of the coastal sea, it might not seem that big. It all depends on actual size and perception.

These first few pages provide us with the basics of "what" this "One To Grow On" is all about.

Two
The Why

Why would mankind want or need to consider the development of construction methods such as the ones that we are discussing? The best answer, short and straight to the point, is to prevent the deep worldwide economic depression and the chaos that would follow it. Imagined as hell on earth with no way to recover. And to make life more pleasant, rewarding, and enjoyable

That is as clear as I can make it. That is the bottom line. The writer's search for reasons for this type of development found many that can be stated a variety of ways. But when we dig for the bottom line of all of those various reasons, almost all of them point to that statement.

This chapter attempts to provide us with real reasons for the need to consider this type of expansion and construction. The bottom line statement sums it up almost completely, with the least amount of words.

There is an old saying that goes something like this, "History is the best teacher." Well, sometimes it is true and sometimes history doesn't apply to the problem that we are dealing with. When we apply the lessons of history in our search to find reasons for considering the construction and expansion method that we are discussing, history does apply and gives us some good stuff to work with. If we look back into history we need to be careful and look for the good and positive parts that apply to the project at hand. If we get hung up on the unimportant, negative and bad parts, it can lead us astray and even cause us to lose sight of our objective.

One of the most important parts of history that helped the writer develop these stories was the era classified as the Industrial Revolution. It was (and still is) one of the most amazing time periods in the history of mankind! History tells us that man scratched around in the dirt for millions of years, before he built his first machine. Then, starting with man's written history from about five thousand yeas ago, things really started to change. The Industrial Revolution started with the steam engine around the 1700s. That is only about three hundred years ago! This would be a good time to sit and ponder for a while, take a break, and consider how far man has advanced; and where he is headed. It only took about 300 years to advance from the exploration of earth in leaky, wooden, wind and oar-powered boats, until man was exploring space and walking on the moon—after scratching in the dirt for millions of years. Yes, it is a great period in the time of man and it started with the Industrial Revolution that gave us machinery and equipment.

There are some pretty amazing changes and good lessons from this portion of man's history. One of the tricks to using history is looking back, keeping your focus, grabbing the good stuff you need and run with it—don't get stuck on or in past things that can derail you and waste your time. Here in 2007, our advances in science and technology, in a way, are products of a continuing and neverending portion of that same Industrial Revolution that started round the 1700s. Let's keep all of this in mind and look into the future; one point of view wants to tell us that there is almost nothing man cannot do; almost anything is possible!

Our industrial history over the past 300 years almost screams at us saying that man can or will be able to build anything that he sets his mind to. The world's engineers have proven this to be true, time after time. This tells us that the type of construction we are discussing can be developed. When man decides to take a shot at it, it is going to happen, one way or the other. No ifs, no buts, no maybes—he will make it work!

Through all of these years of advancement, invention and growth at the beginning of each new project come the words, How do we pay for it? The fact is, every creation, invention, expansion/construction project, and war received the necessary funding to make it happen. One way or the other man begs, borrows, steals, or raises taxes to pay for all of the things that have ever been invented, developed and built—every single one. When we take a broad view of man's past history, it's pretty easy to find a way to agree with that thought. But when we narrow our focus to right now and this method, with money managers pounding their fists on the table asking how do we pay for this, we need solutions. So we sneak back into the library of history and find that profit, demand, and time hold most of the keys of all financial matters. Because of the extremely high and long term cost associated with the type of projects we are dealing with, time and future value seem to be the prime factors of making this succeed. Most important is the time factor and man's ability to make it variable—in relation to cost payment periods, profit margin return etc. History proves that overall future values always increase, with very few exceptions. This is probably the key to making this thing be successful. Another part of the story discusses this a little further.

History almost hollers at us to tell us that man has created value and made money that produced or produces economic supply with almost every single invention that has ever been dreamed up and put into production! This implies a few more things. First, it indicates that all man really needs to do is dream up an idea and make an honest effort to invent, design or build something—sooner or later it will succeed and continuously evolve from that point on. Second, this infers that if man is trying to create economic supply, the larger and more expensive the idea, the more it will produce; creating economic supply and stability in direct proportions to cost. The bottom line says, build it big.

But it looks like there is always a catch with each invention/product that goes into production. At some point during the time of development or production of a new creation, it changes from the choice to make or produce it into the demand that it must be produced—because of all of the financial and economic attachments. As often happens, once you start building something, there is no turning back; it must be built. Simple examples are the automobile, airplanes, and electric generators. Take a look around you—the numbers of these "must be built inventions," is endless. We have every reason to believe that if and when these structures in the coastal seas start, they will follow that same principal and demand to be produced.

Economics is one of the words that this book is built around. To realize why it is so important, we need to discuss a few things. Our dictionaries provide us with a number of definitions for the word. One of the base definitions tells us that economics is the knowledge of goods and services. We can accept that as a definition, but it is not good enough. To get an idea of just how important, vital, and even critical this word is in relation to these stories, we need to add a few pages.

In the world today, economics can almost be defined as a life force. As air, food and water, economics is critical to human life. Said another way, economic supply makes it possible for us to be alive. Economics is the money that we take to the grocery store. Economics is the grocery store. Economics gets the seeds into the ground, the milk out of the cows, the food processed and placed on the grocery store shelves. It is going to work to earn a paycheck. Economics is all of this and so much more.

History can teach us quite a few things about economics, and economic supply and demand factors. We can study the cause and effect at any point in time, and the range covers everything—including international trade, space exploration, and the Industrial Revolution. We can even go all the way back to prehistoric man. Generally speaking, we don't give much thought to the comparison of or between economics and prehistoric man but man and economics kind of grew up together. Prehistoric man couldn't even grunt the word economics—yet it started and developed with man.

Economics started with the caveman, the earth, nature and the spirit. The earth was man's grocery and supply store. It gave him everything he needed to produce you, me, and everything that has come and gone; and that exists today. The big ticket and high value items of those days would have been animals for food and skins, a good spear, a clean water hole, a warm cave, maybe a few fruit and nut trees, and a field of grass, to name a few. We can almost imagine that one of the very first economic booms of those days started with the discovery of flint rock that could make sparks and start fires. The whole thing started with earth and man—everything has developed from and centered on the natural resources and man's ability to think, create, invent, build and manufacture. This basic process has remained constant from then until now and from all indications it will also take man into and through the future. Making good use of, and developing the natural resources that are available, is still the core of man's economic supply and demand systems.

Most of us have expert knowledge of the economic supply and demand systems that are important in our lives. We know for a fact that it is demanded we get up and go to work and bust our butts for eight to sixteen hours a day to earn a paycheck. A paycheck that, for all intents and purposes, is pretty much spent before we get it.

From survival to prosperity, history gives us all kinds of proof about how at the core of man's success, advancements, and economics for thousands of years, we find the natural resources of the earth. This fact should not be taken lightly or for granted. This implies, but does not state as fact, that in order to continue his advance and success he must continue to develop, use, or consume the resources available to him. Right away we can find all types of things for discussion, and to argue about. Things that relate to cost, availability, regulations, environmental concerns and on and on. The fact is, the natural resources provided for, or made possible, all of the great wealth and value of the world that man has today. Farm land, timber, oil, metal, coal, animals and water—these things made the world grow, supplied the foods and made the first millionaires. But the low-cost easy pickings are about finished—except for one. That one is "Rocks"!

Rocks make concrete. Specific rock mixtures and clay mud make cement. Rocks provide all of the metal on earth. The crust of the earth is about fifty miles thick and it is made of . . . rocks! It is the most abundant resource. That makes it one of the least expensive resources. That implies, indicates, and tries to tell us that it can produce the most profit if man can design ways to consume and process massive quantities of it.

Man has the knowledge and ability to mine the crust of the earth and move dirt from now until the end of time, and with very little surface disturbance. The expansion and construction method that we are discussing requires the ability to mine the crust of the earth and consume/process massive amounts of material. History shows us when we have ability, tools and materials, they are going to be put to use at some point in time. Somebody is going to work, and somebody is going to make profit, money, and economic supply. Man does not develop the abilities to do such things to have them go to waste and then permit economic supply systems to fall into economic decline, recession and depression—at least most do not do so willingly or without a fight.

As we discuss reasons for man to consider this, we arrive at the topic of increasing population. This, in and of itself, is one of the primary reasons. In our discussion with this part of the story, as with most of it, we are not going to attempt to get into detailed or factual specifics. We can mention a few ballpark numbers and general thoughts to keep in mind while we go through the story and consider the idea.

History shows us that populations continuously double every so many years. The most recent census of the U.S. population is around 280 to 285 million people. Projected growth rates vary considerably. One of the more common rate estimates shows about a one percent increase going into the future. Some estimates are much higher. At one percent the population would double in about 100 years. If it doubles in fifty years, it happens that much sooner.

Every time the population doubles, all of the factors related to it change. We can consider a simple statement that wants to tell us everything increases—except the ability to produce the economic supply.

Statistics, estimates, projections, predictions and guesses that involve total population numbers, along with all of the variables and unknowns, employ thousands of trained professionals with the skills and knowledge to discuss such matters in a serious and meaningful way. This writer can only mention a few things for thought. The obvious evidence shows, and we can state with some fact and certainty, that the business and government leaders are going to have their hands full working to provide full employment, housing, and food each time the population doubles in size. I hate the thought that future generations might need to be warehoused in high rise apartment buildings without the opportunity to pursue their ambitions and develop their abilities; this is another of the writer's concerns.

The land of milk, honey, freedom and opportunity—that is a place called the United States. Have you ever seen so much milk and honey? Every grocery store in the country has it on the shelves. This hasn't always been the case. It was only about fifty years ago that some of the grocery stores were far and few—with some pretty rough roads in-between, and sometimes the shelves were pretty bare.

Invention, manufacturing, construction, good management, the good portions of government, and hard work built all of those grocery stores. They did not come easy. We and future man can be sure and almost guaranteed that if the leaders of business and government do not stay on top of the economic supply, demand, and balance factors—with real, honest, legal, large-scale construction and business expansion—that the milk, honey, stores and opportunity will go away as fast or faster than they were made possible. Especially when each time the population doubles.

Man's inventions and designs for making money or economic supply covers the range of everything that we can imagine, and more. The largest percentage or biggest part of his designs are honest, legal, aboveboard, and good for mankind. But there is also a percentage of his design that is just the opposite of this—some of it is very nasty, dirty, illegal and immoral business. This is nothing new; the problem is as old as man. The real problem is the percentage of it that the core of primary business depends on, relies on, or needs—in the process of making the profit required to keep the economic systems functioning and afloat. We can think of this section of the economic systems as one that only makes money temporarily by moving money or funding from one place to another by illegal means. It is the constant battle between right and wrong. This places the duty or demand onto the side of right to invent and design ways to create honest and legal methods of making the economic systems function—so the need for the illegal, corrupt, and immoral portions can be reduced, eliminated, and controlled. The best way to do this is to invent and build new "stuff"—maybe the construction of city structures that rise up out of the coastal seas will help.

Economic influence and computerized electron control. The old caveman's offspring really stumbled onto one, when they discovered how to control sub-atomic electrons with a particle of silicon sand (rock fragments!). I think it was about 1947 when the boys invented the transistor—a small component that can control the way electrons move in a low voltage electrical circuit. This one little device, or gadget, led the way to all of the electronic stuff that exists today, and will exist in the future.

Great things have occurred with the invention and evolution of this small electron switch; it has helped to create a lot of economic supply—we might be able to say, hundreds of trillions of dollars. That's a lot of bucks in fifty-some-odd years. After it brought us things like transistor radios, color television, radar, space flight, and all of our other stuff, it evolved into the microprocessor that runs today's computer systems, beginning twenty to thirty years ago. Now if we compare development processes of the past to those that we can foresee and imagine going into the future, stop and look at how things actually are at the present—we can realize that our high speed electronic control and computing systems are still in their beginning or infant stages of development and evolution. This indicates, and almost proves, that only a very small portion of the power of the machines and their influence on the economic systems of the world has been realized so far—both good and bad.

Man has come a long way with his designs for machinery and equipment. He now has fully automated computerized systems that purchase materials from other computers, then supply the materials to other computes that assemble new computers, then are sold to other computers, with payments made to other computers. He will need to be careful going into the future so that he doesn't work himself out of a job and end up standing in a computerized soup kitchen waiting for some digitally processed soup.

A little sarcasm can be good sometimes, but it does illustrate that man can piggy-back automatic systems that run themselves. Such system designs are reaching the point that thousands and perhaps millions of additional jobs performed by people could be replaced by some of man's machinery. In some cases it would be ok, in other cases, no so good. Now that he has perfected his machinery, another one of his creations has come back to bite him on the butt every once in awhile.

He might discover that the next challenge is designing projects that can bypass a few computers so that he can create paychecks and supply the tax base. An opening for this type of project design might be large scale construction in the coastal seas.

When we permit our thoughts to wander off into the area of science fiction, keep in mind the science fiction of the past has developed into "science fact." Study man's most recent technological advances; our imaginations can run wild, if we have time to permit them to do so. One point of view that has wandered into these areas has discovered quite a bit of interesting stuff. We don't have time to discuss much of it at the moment but we can discuss some that applies to our thoughts of man's future economic needs. With a little study and imagination we find that man can design and build machines and systems using high speed computing capabilities and radio systems that can do endless numbers of things. We can imagine and might know of technological advances that can reduce costs, save time, save lives, provide safety, prevent destruction and make life easier—but they can't be developed because of the negative economic impact that would result from doing so. We can think of it as technology that must be withheld or put on the shelf until some other way to balance the economic impact is invented or devised.

As a really far-fetched example, let's try this: imagine that a couple of the mad scientists had finally discovered the one magic pill, the one compound that could cure anything and eliminate all of the other medicines. Such a discovery could be a great thing for mankind. But on the other hand, it might have a negative or disastrous economic effect—unless man could devise a way to make up for the economic losses.

Here is another one. Suppose that a couple of the nutty professors had finally discovered a way to control the weather. They could make it rain or stop the rain; they could make the wind blow, or not; could stop tornados and hurricanes—but the economic impact from rebuilding from such naturally occurring events would not permit the technology to develop until an economic replacement was found. The technology, for examples comparable to these, must be withheld until the time that economic replacements are developed. Such economic replacements might be required to generate or produce hundreds of billions of dollars annually; new inventions that can provide this type of economic supply are far and few between.

Then someplace along the road and out of the blue, comes a story from a couple of old economists. Now these two old-timers have been at this business for quite sometime and they, in one form or the other, will be at it till the end of time. We might think of them as existing someplace between science fiction and spirit; and hooked up in ways that are beyond explanation. Regardless of how we might think of these two ghost writing economic experts, the story is the important part.

Now these two old duffers have seen it all. They know the ins and outs of the way things work. This idea of construction in the coastal seas, involving massive amounts of materials, expense/cost, profit, economic supply, and personnel is a solution for mistakes of past economic designs; sort of righting the wrongs of the past. They know that to transfer large quantities of economic supply source, from parasitic and lawyer infested design methods that are in use today, it will take extremely large alternative source supplies. Along these same lines, if man is to be permitted to introduce all of the time, cost, and life saving technology that is in storage and on the shelf, make way for all that will develop in the future, it is going to take that many more billions in annual economic revenue. The idea of expansion into the coastal seas with continuous construction, for all intents and purposes, is unlimited—it can accept all of the money that is put into it, use it, and produce a profit. This indicates anytime one of the primary economic supply sources starts to drop off, it can provide the alternative or replacement supply.

As an example, let's use the automobile industry. It is on the verge of major change—high speed magnetic levitation trains seem to be in the not too distant future; cars are more efficient, they last longer and new alternative fuel designs are in the process. Anytime the automotive industry declines, in any way, the economy suffers. Any portion of economic supply that is lost in the auto industry must be made up someplace else. After long-term development and use, the ocean construction process should be able to use all of the equipment that needs to be produced. Well, this piece mixes a little fiction, science and fact to demonstrate why this needs to be considered.

There is another old saying from the days of the past

that goes something like this: "If it looks like a duck, and it quacks like a duck, and it walks like a duck—then there is a pretty good chance, we are in fact, dealing with a duck." Now along these same lines, some of the old-timers claimed that there was a sixth sense that permitted them to smell money that was about to be made. It was also noted as being able to sniff out a moneymaking deal or proposal. Well, like everything else, we can figure that sometimes it works and sometimes it doesn't. We can add a few twists to this line of thought and come up with one of our own that goes something like this: if it sounds like a moneymaker, and it has the overall appearance of a moneymaker, and old money bags comes over and gives it a smell and a thumbs-up—then there is a pretty good chance this thing is going to make money. It used to be that in cases such as these, the old boys would be going after ten, thirty, or fifty percent of anything that sounded like construction in the sea.

Sounds too good to be true; there must be a catch (we might say to ourselves). There are a few, but they do not amount to much—compared to the economy caving in. Environmentalists, ocean rights protests, and fishermen are the first that come to mind. Then we get into property rights, jurisdiction rights, and international water and boundary rights. The bottom line wants to tell us that all this really means is that the project will provide work for all fields, even the lawyers and politicians. One thing to remember about objections or attempts to block it—if the economic systems fail, rights and lines won't mean much anyway. Importantly, there is a chance the development of the methods are in preparation for something vital.

In the U.S., we have millions of people who live paycheck to paycheck—one that is spent before received. People who barely get by, and the many who don't. If man is ever going to reverse this trend, and make life easier, it will require something that is real big. When we get into the areas of international trade and dependency, combined with consolidated, efficient, streamlined on-time production, delivery and distribution, it does not leave much room for disruption of process, without major destructive side effects. Any major disruption within the combined systems can and often does have impact all around the world. One major event can put millions of people out of work; when that happens, life savings can be used up; debt increases; bankruptcies destroy families and lives—many people never recover from events such as these. One of the primary reasons is because of the debt load. If man could do one thing to provide insurance against this, and at the same time lighten the load on all people, it would be to provide them with a little more money to reduce the debt. This in turn can provide the cushion, a little leeway, or a means to absorb the problems and unexpected events that cannot be avoided. This can reduce the destructive effects all the way from the individual to the top of the corporations, providing savings, with more stability and growth potential.

At this point many of us might be thinking something like: ok, those words sound good, but it's all a bunch of blah, blah, blah—we all know that we could use more money, tell us something new. Stick with me for a while, we are going to take a look at designing pay raises—from the bottom to the top, but it is going to take a few pages.

The economic systems and debt. It is amazing how the economic systems have evolved with time. Most notedly, monetary value. Next are the methods and principles of doing business. It was not very many years back that people purchased and acquired things, only by either paying cash or making a trade of some sort. There were no credit card systems, and bank loans were a very serious matter. One of the advantages to that way of doing business was that individuals, businesses, local, state and the federal governments carried little or no debt; budgets were forced to balance—it was pay as you go. On the other hand, many people had to make-do without the luxuries and necessities of the time. This goes back to the time period when wages of five dollars a day was big money. But our parents and all of the parents before them made the systems, made them work, and here we are.

The increasing use of debt has created large changes; some of them are really good and some are not good at all. If we study the evolution of the process of making the economic systems function by using debt and credit as opposed to balanced budgets, we can learn a few things. From a distant viewpoint things look pretty good, overall. When we look a little closer, there is reason for great concern; due to the amount of debt burden, from the bottom to the top.

Conventional thought and wisdom want to tell us the amount of debt carried by individuals, families, small business, corporate giants, and all governments needs to be reduced—because a large percentage of individuals, businesses and governments cannot pay their debts and are very close to being bankrupt.

Now this next part has some fiction mixed in with the nonfiction—it mentions things like pulling rabbits out of hats and makes an example that uses a fictional boss. It's kind of like this: the boss would like to give everybody a substantial wage increase, but we permitted our cost reduction and efficiency experts to take things a little too far during the process of cost reduction, streamlining, and consolidation. Over a period of time, all of the cost-cutting measures ended up actually reducing profit margins to the point that a large percentage of the business is one major screw-up or event from bankruptcy. The problem is, if the business does not receive some large orders soon, the manufacturing plants will need to be closed. We might be able to reopen them in the future or they may be closed permanently.

History proves that business managers all across the country have had to say words to this effect, in the past. That indicates that they are going to be repeated in the future. Anytime any of the major or primary industries announce that factories or manufacturing plants will be closing, it is an indication of serious financial and economic problems. Anytime one of the automobile manufacturers announce problems or plant closures, there is big trouble and it is not very far away.

Sometimes the boys can pull the old rabbit out of the hat and turn things around with a financial miracle, and sometimes they can't. When they can't, history shows that the business cycle goes down, finds a bottom point, and then starts back up again. That was in the past, and big changes have occurred since then.

Some of the changes of the past twenty-five years or so, have to do with repayment of debt, domestic manufacturing; heavy primary industry—that includes high quantity domestic production of the natural resources, along with the value created by manufacture through each process, until a finished product is actually consumed or put into use.

World War Two and the effects of that war built and still provide the basis of the worldwide economic system that exists today. After the war, the world had a continuous economic boom, with up and down business cycle swings that were directly related to the heavy primary industries, and the use of natural resources. In the United States, for something like thirty years after the war, the country was under construction. New construction of all types and lots of it. There were multiple high rise building projects in numerous cities across the country; the interstate highway system was designed and under construction all across the country; many parts of industry ran at maximum capacity; natural resources were low cost; and every time a rock moved in the process to make steel, concrete, chemicals, or medicine, it was making money at each step of the process! Most importantly, the value added by manufacture during each of the processes was making money. Slumps and downturns in the business cycle, for the most part, didn't last long. Many of the factories, mines and projects couldn't hire enough people and jobs were plentiful. Military defense demands required large quantities of just about everything, especially steel. Great amounts of real, solid value that will last until the end of time were created during those thirty years.

We can generalize and say this economic boom, that ran on heavy industry and construction, started winding down around 1980 and a new set of changes started to take place. When a "boom," expansion, or growth period starts to decline—economic recession is next in line. Some time between the recession and depression period of the 1980s the electronic computer industry combined with civilian access to the Internet communication system; then it evolved into an economic generator that created enough economic supply to help keep the economic systems growing. We could say that an economic boom developed with the computer network, but compared to booms of the past, it made a large pop rather than a boom. Great amounts of economic supply are, and have been, generated and managed with the new technology. But private, corporate and government debt has continued to increase. One viewpoint wants to tell us, when real, solid value and wealth of a nation is in the process of being created, debt is low, the bills get paid and budgets balance. The same view indicates the opposite is true. This is important, because it is a very simple indication that in order to pull everybody out of debt, get the bills paid, and budgets closer to balance man will need to build some extremely large projects.

As the current conflicts and the war on terrorism start to wind down, so will the economic supply it generates. History tells us and indicates that it will be the beginning of something else that will provide for the necessary economic needs. Some of the analysts predict more terrorism problems. Another viewpoint says that the rag-tag terrorist groups don't have much chance of causing large amounts of trouble, now that security is in the process of being restored and updated. As in the past, this adds up and indicates a few other options that could develop to produce expansion, growth and an economy. Advances in technology and new inventions have created large portions of the peacetime booms of the past; for most of us, this is the preferred way to grow the economic supply systems.

We know for a fact that some of the science fiction of the past has developed into science fact of the present—that it exists and is right here, right now. It's time to get creative. Let's put a twist on science fiction and try one called economic fiction. First thought, economic fact proves that man has problems, such as being buried in debt, and barely able to pay the bills. Economic fiction tells us the same way science fiction tells us, man can do almost anything! Suppose man is getting sick and tired of being sick and tired. He is getting old, and he is tired of breaking his back, working eight to sixteen hours a day. He wants to spend a little time with the wife and kids, get out of debt, make some money—take long extended vacations and enjoy life. That's not too much to want out of life. Economic fiction tells us that man can design and engineer an economic system of good, peaceful, filthy rich profit that can spread around the world. The fact of the matter is, that it is already in progress and all that man needs to do is refine, correct problems, and build stuff.

Well, we all know how man is; he gets soft, spoiled, lazy and is never content—always wanting more. He started out sleeping on a nice soft mattress of leaves and pine needles; he went to feather beds, cotton and spring steel beds, air beds, water beds and sleeping floating in outer space—as if to say, almost nothing is ever good enough and everything needs to advance and improve. If man is going to pursue and rewrite the economic fiction of 2007, he might as well take it just as far as he can. History proves the past designs that continually increase value, double wages and costs, and cause inflation, are successful. We could say that some of the success was due to careful and educated planning, some of it was of the economic fiction classification that became fact; some of it was just plain dumb luck, and that most of it was probably related to the grace of God. Regardless of how we perceive the success of man's economic designs, the designs just aren't working as well as they should.

If we wander off into the area of scientific economic fiction and look for solutions that are at least in the neighborhood of being real, someplace in there we find that production, sales, wages and taxes need to increase, while at the same time, holding costs and expenses. Fact and fiction combined, almost prove there is a formula that business and government can agree upon, that could permit production, sales and wage increases—if windfall profits were time delayed and used to restructure the economic and tax systems.

Try to imagine it this way. We have a manufacturing company that is operating by the skin of its teeth. Now suppose old money bags comes over with a contract, a contract that is different. It will require full production from the plant for a number of years. But, there is a catch. His contract says that the old man is going to pay for the cost of the product; current plant expenses with no major increases or modifications; plus wages, salaries and benefits. But all of the profit produced above the current level goes to a special fund that is going to be used to double salaries and wages of the employees. The next part of the contract increases production by 100% and control of the plant and profit return to the owner—with provisions that keep salaries and wages at the new increased level or above. We can add a little more fiction that says the old man has a contract that covers about fifty million people and includes the same type of requirements in his plan to raise cities up in the coastal seas. We should wonder and study the possibility of man, business and governments being able to make such things happen.

We could fill books with thoughts and talk, and talk about more talk—about why man should want to consider an idea of construction and expansion such as the one we are discussing. But, like the old boys used to say, "Talk is cheap and a lot of hot air won't get the job done." The only thing that will get this type of construction started is the money that some of the investors are willing to put into it. The bottom line is pretty simple really; future man, woman and their kids will need a place to grow, build and live—they will need one to grow on.

* * *

We have a couple of old sayings that apply to pictures. Sometimes these old sayings seem to have meaning or significance and sometimes they are just old sayings. It all depends upon the situation and the time. Anyway, one tells us that a picture is worth a thousand words. Most of us can relate to and understand that—especially if we need or want to. But most pictures arc just an image that is produced onto something representing something else. The other old saying that I am thinking of goes something like this, WHAT DO I HAVE TO DO, DRAW YOU A !@#$%^&*() PICTURE?!!!!!

This is one of those old sayings that we don't get to hear much anymore. We could most often hear this line around the factory and shop areas when a supervisor was attempting to get some task accomplished by some, less than willing, coworkers. Although it has been used or at least thought of in many other locations and situations.

As part of the process of attempting to explain parts of the construction method that we are discussing, we have a couple of pages called the vision an the proof. It's probably a few hundred words that describe an imaginary photograph or picture, a place, a time and a few thoughts. This is one of those times to use your imagination to think about the idea.

The vision. Picture yourself on a large pier that is built above the surface of the sea. The location of this imaginary pier is at a coastal resort town. It's a nice day; clear, warm, and calm. There might be some kids building sand castles on the beach; a few swimmers, sunbathers, fishermen. There could even be a few boats off shore, just a nice day with a clear blue sky. The pier we want to try and imagine ourselves on might be one of the older wooden types—with a wood plank deck and vertical support posts that are driven into the ground, or it might be one of the newer type made with concrete and steel. Let's try to imagine one of these newer ones made from concrete and steel, and focus on that. Try to imagine this pier as solid, strong, heavy and secure. It could be 50 to 100 feet above the water. If you could stomp your feet on it, it would be strong and solid. We should imagine this one as being a long one, it goes way out there; a thousand feet, maybe a quarter of a mile—and wide, 150 maybe 200 feet wide. It's a big one! Now, this is one of those piers that has a gift shop, a small restaurant, a bar and a couple of other small shops. There might be a kid there that rents fishing equipment; could be a few people sitting around on benches, a few fishing, some just walking around. To go out past the shops toward the end of the pier you might have to pay fifty cents or so.

Imagine that no one is out at the end of the pier and we are going to stroll out there and discuss this thing. If you are daydreaming, like I ask you to, it's a real nice day, no problems. Looking around at the end of the pier we see calm blue seas, and a clear blue sky. Take a look back toward shore—take a look up and down the coast. Not bad. Now, turn around and take a look out over the ocean—way out there where they sky and the ocean meet at the horizon. Here is something interesting. One of the formulas for the curvature of the earth tells us that it is about two and a half miles out to the horizon—that place where the sky and the sea meet and the sea disappears over the hill. This tells us something else. Suppose we had a structure that was 450 feet tall (measuring from the surface of the water up), if it was placed a couple of feet above the water, out there about nine miles, we might be able to see the top few feet—if at all. This tells us man could go out there a few miles, build an entire city, and never see it from shore. Something to keep in mind, if you are interested.

This next part stretches the limits of the imagination; let's take a shot at it. Sit yourself down on the bench and take a look toward shore. You are looking into the past, in a sense. If you can imagine looking back through time, 700 years back, all you would see would be forests, animals and a few of the native Indians; all the way to the other coast—border to border. Take some time, use your knowledge of history and try to imagine how the country developed over the past 500 years, bringing us to 2007. Focus on the good and important things—not the wars and all of the shit. The growth and construction, the people, the advances in knowledge and science. Remember that it was about 1900, only about 100 years back, that things really started to happen—the first airplane, electric lights and power, cars and trucks. If we could describe the past 100 years with one word, *miraculous* says it about the best.

The last paragraph says volumes; if you daydream a little, you can think of more than I can put into words (as is the case throughout the book). Points to keep in mind are the rates of growth, population, types of development, changes; and note that the population of the U.S. is reported as doubling in the last fifty years. Some view from the pier, huh?

The proof. With this story, we are discussing a new construction and expansion method that might be in man's future. The pier talked about in these few pages, and the piers that actually exist, provide us with the most basic factual proof that construction above the sea has started—it does exist, it is real, and it is successful. It is very primitive with respect to building a city on a structure that rises up from the sea floor; but it is real and it is the starting point.

One point of view can tell us it is kind of like the comparison between the Wright Brothers airplane and the space shuttle. Think about all of the "stuff" that was manufactured and produced, worldwide, that relates to those first small wooden airplanes and today; study the economic supply that has been produced. It was only something like seventy years between the small wooden airplane at Kitty Hawk, until the space shuttle was developed and flying around the Earth!!!

It is almost demanded that we wonder and think about today's piers developing or evolving in a comparable or similar manner.

* * *

It seems that all through life, there is always more to the story than meets the eye. As though there is always something; on the other hand, the other side of the coin, when the other shoe drops, the other side of the mountain, around the next bend, the next in the logical line of progression, the action and reaction factors, the consequence of actions etc., etc. Well, the idea of the development of the type of construction that we are discussing is no different. Let's get to the point and then we can add all of the et ceteras later. This has to do with the fact that the sea level rises due to the melting of the ice at the polar caps of the earth. It has been a slow process over the years. But the indications are it is one of those things that will catch up with man, sometime in the future, and create very serious problems that will demand solutions and actions. Although a fifty-foot rise of the current ocean level could have devastating effects for man, the fact the threat exists might permit it to be classified as a blessing in disguise, due to man's need for economic supply.

Before we go any further let's make something clear. This is a layman's work of personal opinion. There are experts who study all of the things we are discussing, who have accurate facts, estimates, and predictions for determining required courses of action needed to protect and plan for most any problem man will come up against. But a book such as this permits us to suppose, wonder and maybe realize something that might have slipped by.

History provides a lot of good information about the ice ages of time. It tells us the melting of the ice that covers the earth is a slow process. It also shows that the last ice age ended eleven or twelve thousand years ago; so far, the melting ice has added something like 300 feet of water to the level of the seas. In a time period of man that is so concerned with fractions of a second, microseconds, and nanoseconds, facts that pertain to thousand-year periods might seem to have little or no relevance, but they do.

If we attempt to simplify this, we can form some ideas with a few numbers. History shows that the last ice age ended around 11.5 thousand years ago; other studies show the sea level has risen about 300 feet since then. If we work the numbers, they indicate sea level increased about .026 feet per year; that works out to something like .313 inches per year. We can take it a little further and they show a rise of one foot about every 38.3 years. This tells us the Atlantic Ocean, if left unchecked, would reach the White House lawn in about 957 years or so. That would be around the year 2962. (That is about a 25-foot rise; Washington is about 25 feet above sea level, according to some of the numbers.) This all indicates, within one thousand years, there will be major problems worldwide if man doesn't successfully prepare and install the proper safeguards and structures. Well, we have some variables that affect these numbers. The averages could change considerably from 2005, proceeding into the future. They could be higher or they might be lower, that is for the professionals to figure out. We can think of the average numbers this way. If the averages have changed and indicate the White House lawn won't be ocean-front property for another 2,000 years, that would mean man has a considerable amount of time to rectify the problem. But, on the other hand, if the averages show it will only be 500 years until the problem makes it to Washington, man will need to get seriously concerned about preparing solutions, in a hurry. We could think of it this way: If Washington goes under, a large part of the world will disappear—so governments have the duty of preparing the solutions for the problems far in advance of the time that they will occur. This is not a problem that will affect only the U.S.; this one goes around the world.

For some reason, this part of the story didn't develop until the book was almost completed. Conventional thought says if we are going to prepare a long-winded story, the real important stuff should be up-front. Well, that all depends on our understanding of man; the influence upon the writer; and in general, how things really are. It's kind of like the bear in the woods. As long as he is at a distance and we are not in immediate danger, we let the bear go his way and we go our way. It all has to do with the time factor and how fast methods of development take place. All through history some whacked-out inventor or professor has dreamed up ideas, that over time almost always develop into something necessary or useful at a specific time—or maybe right on time. Now some of us are wondering what does this have to do with the construction method that we are discussing—and the amount of water that is in the sea. It has to do with the development of the process methods, and the ability to manufacture and place massive amounts of concrete and steel, in, near, or around the coastal seas; and money; and time.

The indicators tell us man's dry land, in the coastal sea areas, will be permanently covered with water sometime in the future, if he does not prevent it from happening; or, if his preventive measures fail. Some of the same indicators help tell us man will need a profitable or moneymaking way to develop the preventative processes.

Eliminating all of the wordy details and b.s., the bottom line wants to tell us: in order for man to build seawalls that can hold back rising sea levels, he will need profit-making projects that will permit him to refine and perfect his designs in a timely and orderly manner, so that normal peacetime life can continue to grow and prosper during the process. For instance, man builds his first demonstration project that places a city on a concrete and steel structure in a coastal sea location. With the success of such a project, man will have perfected some of the large scale processes and machinery that will be needed to build seawalls. He should have it made, providing that he does not wait too long! Entire civilizations have been destroyed, and disappeared in the past, because of man's failures of one type or the other. If he has to abandon the coastal cities and run for high ground, we can imagine future man will lose a large part of his civilized development, only to become the object of some future anthropologists' and grave robbers' so-called "dig." So much for the bottom line.

Many of us could be wondering what kind of time periods we should be referencing. It's hard to say; most coastal cities and other areas have problems with big surges and waves that accompany large storm systems now. Erosion is already a problem in many areas. Oftentimes weather broadcasts show homes collapsing into the sea during large storms; it just slowly and continuously gets worse. We can imagine quick fix, short-term methods of breakers, seawalls, truckloads of rock and dirt will continue to be used to take care of the problems. But after a number of years such methods will probably lose their effectiveness. If we use those old average numbers that tell us there was one foot of rise about every thirty-eight years, it gives us something for thought. The writer doesn't have the current average numbers; they could go either way. The current average might be one foot every fifteen years, or it might be one foot every fifty years—we will need to wait for the new numbers.

Some of the indicators want to tell us the rate of rise has increased in recent years. Sometimes reports are released that tell us things about global warming, increased solar flare activity on the sun, and huge icebergs that break away from the cap areas. We can also add the scientific theory that tells us the core of the earth is someplace in the neighborhood of 10,000 degrees F., and that the ring of fire over there in the Pacific, alone, has something like 300 active volcanoes. (There are even a few at the South Pole.) These all want to tell us it is time for man to make the necessary preparations to deal with rising sea levels. Man can pull off some pretty amazing miracles sometimes, but man is not prepared to control the rising temperature of the earth that is caused by the power of the sun. If the climate is starting to heat up a few degrees, most likely it is beyond man's ability to stop it; and he would best spend his time and resources preparing to deal with the effects—such as rising sea-levels and the melting of the polar ice caps.

*   *   *

A blessing in disguise. Maybe I need to explain this one. An important part of the bottom line wants to indicate there is a potential for creating a worldwide public works project to build the seawalls needed to provide protection from increased sea levels; it will be an economic generator that will be demanded by the forces of nature. If man gets his timing, designs and plans right, he can use this trick of nature to create economic gains, security, freedom, independence, work, paychecks and overflowing tax coffers. It is kind of like a one time offer, or a one shot deal offered by God—if man can put the designs together, he might turn the world into the place it was supposed to be. When you start adding up the requirements and costs to construct 100, 200, and 300 foot seawalls, along the coastal areas of each country, it requires a lot of work and a lot of money; that in turn will produce a lot of work and a lot of money. Then, after man gets his seawalls built, he can use the same technology to construct his future cities on structures above the sea. Kind of like a two-for-one blessing in disguise . . . if man can keep from screwing it up.

Let's look at some of these numbers again, to kind of reinforce our understanding. Some of man's studies, estimates, and predictions tell us that there are about 300 feet of additional sea water locked up in the ice at the north and south poles. The old averages indicate that it will take about 11.5 thousand years for all of it to melt. That rate is about a one foot rise every thirty-eight years. In general thought, the impression might be that man could take care of things with current methods—handling another one, two and maybe three feet of coastal sea elevation. But anything over three feet, from one point of view, seems to be about the maximum. The writer doesn't have anything scientific to back that up; it is kind of a combination of instinct and experience. Actually, another three feet could cause very serious problems in many locations—we will have to wait for the reports. Three feet times thirty-eight equals one hundred fourteen years. Man could place a lot of miles of seawall in one hundred fourteen years. But if the rate of rise increases to .62 inches per year; that is about double the old average and makes another three feet in about fifty-eight years.

Well, fifty-eight years, one hundred fourteen years or one hundred fifty-eight years—at any rate, man needs to get started with something that will provide strong, secure, and permanent protection that can continually be increased through the years to handle that additional 300 feet of ice water. The real answers for most of the questions relating to this problem will be found in accurate numbers. In addition, there is no way to rush order this type of construction—man cannot try and delay or keep putting this off. If he fails, or waits too long, future man will not be able to keep up with the problem.

If we sit and think about this for awhile, we come across all kinds of thoughts. One passing thought wants to tell us man could just evacuate the coastal cities and move inland to higher ground. Building new and rebuilding as he goes inland away from the sea. But that wouldn't work. Imagine mass migrations of the city folks from New York and Miami moving into Kansas and Nebraska—it would be like the show *Green Acres* multiplied by millions; it just seems like it wouldn't work out the way it is supposed to.

Let's review for a moment. I can almost hear the question being asked; what are we going to do, build cities above the sea or build seawalls? You !@#%%^*&! The story has taken a different course that changes the meaning and the need, with this chapter. When we study history, even briefly, we can find that all of man's developmental processes were put together or happened in bits and pieces. It's like this: all of man's inventions and industrial developments must pass through certain stages and processes—ones that produce the next crucial step of development. Sort of like a puzzle that keeps adding pieces once we complete the one that is being assembled.

If we look at the history of man's industrial processes, and apply this bits and pieces thing with the new pieces of the puzzle that continuously gets larger, then consider the fact that seawalls will be demanded at some point in the future. Ponder it for a while; we can form a few ideas. One point of view can form an idea that wants to tell us all of man's great industrial power has developed in these bits and pieces, always leading to the next stage of development—it says all of these abilities man has produced are in preparation for the next demand. Man has developed his science to the point that he now has factual proof of the time periods of ice ages that have occurred through history; he has proof of the amount of water produced by all of that ice; he has proof and science to tell him how much water can be produced by the remaining ice; his science and proof tell him that the climate is getting warmer; he has the tides timed to the minute; and he has mapped the sea floor. We add this in with his industrial abilities that produce machinery, steel and concrete; and experience building dams, dikes, locks, levies, ocean vessels; along with all of the other supporting science and knowledge. One conclusion that can be made, tells a man will need to use all of these abilities and knowledge for some task that will be more difficult and larger than any that he has encountered in the past. Reasoning wants to indicate that the types of construction and demand for this type of construction, fit in—because it is larger, more expensive, harder, and more vital than almost anything that has been demanded so far. If we add this type of thinking, together with our discussion of anticipated economic need and combine the part about a blessing in disguise, well, in the writer's opinion and until science proves no, the general idea sounds right; we should hope, and probably pray, that it will work.

With this brief review, we can think back over these first fifty pages and form an idea that can tell us if future man could develop the methods, through a moneymaking process that includes the construction of cities on structures in the coastal seas, before the seawalls will be required, it would also fit in with the process of man's past development. The reasoning behind this is because, again, from a distance and in this point in time man has a choice; it's not demanded that he develop such technology—to some degree we could consider the idea folly. Folly in the same way that putting a man on the moon, before he succeeded in getting there, was folly. Folly in the same way that putting a man on the moon, before he succeeded in getting there, was folly. If future man gets the timing right he could build the little 100 square mile demonstration project in a timely manner, reaping all of the benefits associated with such a project—in preparation for or in conjunction with the planning of the seawall protection systems. We could think of it as developing the necessary technology with a combination of private investment and government investment, into the demonstration project prior to the demand for the seawall systems and costs. Those thought of us being strictly governmental in nature, in the way that the military is. Such a process could really ease, or at least alter the tax burden. Now, with this in mind, the writer intends to present the science fiction story of developing the structures to raise cities above the coastal seas. It would take a few more books to present ideas for constructing seawall protection systems that include lock and lift systems for ships, gigantic pumping systems, and so on.

As we consider all of this "stuff" that we have discussed so far, let's add this into the old memory bank. The earth is reported as having a surface that is about thirty percent dry land and seventy percent water covered. When we start deducting barren desert area, the high rugged mountain area, and the extremely cold and hot climate areas, we begin to realize a few things: when we consider our coastal areas we find they contain the largest percentages of the population; along with this we find a large percentage of the world's oldest and major cities were built next to the sea. This in turn tells us it contains some of the most valuable and most productive land on earth. Then we can also add that some of the most fertile farming regions are located within these areas. This says if future man permits the sea to rob him of the coastal area land, he will lose a large percentage of the best dry land that is available on earth.

If man fails to control the problems that are indicated, we can imagine a large percentage of the civilizations and cultures will disappear, only to exist in the scraps of history someplace in the future. We might also imagine such a destructive process could be worse than any such event of man's history, and could even be compared with events such as large scale nuclear destruction. If we dwell on it long enough to realize some of the possibilities.

Here in 2007 we can be fairly sure that all of these potential problems and the solutions are being studied from all the different angles by governments, businesses, universities, theologians, and individuals worldwide. All of who, will, have, or are producing a wide range of thought and fact related to the problems and the necessary solutions. Most of us are consumed with the events of right here and right now; trying to do our jobs, provide for our families, get the kids educated, get ourselves educated, or are simply trying to hold on for a few more years. While in other parts of the world the wars of the old hate still attempt to kill as many as possible. Regardless of the category that we fall into, most of us learn of the major world events through the various news sources; and we learn of events and problems, such as these, when they become headline material. This type of thing works the other way also. Long before problems make it to the headline news, often the solution was developed or planned years in advance. It has something to do with the professionals and experts that manage and take care of stuff.

One of the things the reader needs to understand is that the writer wanders off to areas that do not provide the latest news; it could be that he has missed the reports explaining the most recent solutions or developments. But, just in case, we need to keep in mind time factors that apply to the types of things we are discussing. We have heard some of the old references to time: how time waits for no one, time slips away, and others. Well, time is devious and sneaky! When we are young and, like the old timers used to say, full of piss and vinegar, time can drag on as though all of those big events of life will never take place. Then, within a matter of years, things change and ten years can slip by, leaving us to wonder where the years have gone.

When we reference time, with the construction methods we are discussing and the polar ice caps melting, time becomes very important. The reference to time waits for no one will come into play if the dry land of the coastal areas of the world should start to flood beyond control. It's like this: the indicators tell us man has a certain number of years to prepare in advance of the ice melting; before the point in time when the coastal cities will need to be protected or evacuated. At the point in time when the water would wash over and flood the underground utility and transportation systems, that would be close to the final call to evacuate. If we reference this to New York, Miami, New Orleans, Los Angeles, and Seattle, just to name a few, we can see how time will not wait for anyone—and the problem could hit all at once.

If we take this type of thought a step further, we can see from one point of view: if man fails to construct the necessary protective systems and permits even one of the major cities to require evacuation, it can disrupt all of man's other processes that are interconnected around the world. We might be able to think of it as a destructive chain reaction, that will have to run its course if man is not prepared in advance. On the other hand and opposite to the doom and gloom, if man can perfect his engineering methods and processes, and get the timing right, he can turn this problem upside down and take full advantage of it to drive and supply his economic systems—for maybe thousands of years.

Let's take this just a little further using an example that uses numbers; they are not accurate numbers, they are for thought and discussion only. Suppose that studies between some of the world's best engineers came to a conclusion that this problem with the rising sea levels is going to cause serious problems. Suppose the studies show man has about 150 years before major protective efforts and construction must be in place; and at 200 years, calculations show that things that get real serious at many locations. The same studies might show temporary quick fix projects and prevention measures will hold most of the problems at bay for another fifty years or so; also included in this fictional study are the calculations that show, with major efforts, the primary seawall, lock, and pump systems could be constructed and in place in about 100 years, with plenty of cushion for changes and unexpected events. Many of the world's smaller, low elevation islands, however, might be lost to the sea.

In this hypothetical (long shot) example, mankind might have it made. We know for a fact man can do and build amazing things in the span of 150 years. In this brief fictional outline, it gives man fifty years to develop and perfect his methods and processes, and then 100 years to get the primary work completed. Engineered and timed properly, it might have a chance at creating all of the benefits that we have discussed, and more.

As we continue with this fictional example, let's consider the thought of a fifty year development period. This amount of time can produce extraordinary things. For example; it took about forty-five years to develop the rockets and knowledge to fly a man to the moon and bring him back; and get this, most of it was done with slide rulers and the early calculators. Well, to put it simply, man can develop things in much less time with all of his added advancements since then. The history of the U.S. tells us a large portion of the country's business and industry has started and developed with private investment laying the groundwork for most of the primary industries that exist today. It is a time tested and successful way of doing business and financing the government. These projects we are discussing might, and hopefully will, be permitted to develop within the same successful guidelines, if and when needed.

Now let's condense all of this into one paragraph that sums it up, states the idea, and answers the question: are we building seawalls or cities above the oceans? The writer made an attempt to interpret some history, some science, and some fiction. A story started to develop; it discusses building manmade islands and cities that rise from the sea floor in the coastal waters of the world. Then, the ideas that relate to rising sea-levels developed. This led to ideas of the construction of massive amounts of seawall and lock structures. The study of history, in preparation for the future, indicates man will be permitted to develop the necessary processes and equipment for constructing seawalls, by starting developments of the manmade islands and city structures years in advance of when the seawalls could be demanded. The indicators want to tell us the whole combined process might help supply large portions of the economic demands for the world—for hundreds, maybe thousands of years. It is interesting to note this idea, combined with fact, about the rising sea-level, is an event in man's developmental process that will demand worldwide cooperation, or everyone loses. But please remember, this is an independent study and was not professionally prepared.

Three
The How

You propose to do what?!!!!! You must be out of your mind, building cities above the sea! Just how do you propose to do it? It's impossible? You want to build a spaceship and fly it to the moon? You must be nuts! That's impossible! They should lock you up and throw away the key! Ha, ha, ha. Just how do you propose to do it? A machine that flies in the sky? If man was supposed to fly, God would have given him wings! A boat that runs—under the water? You must be crazy! Buildings that are 500 feet tall! Ha. Internet computer systems? Can't be done! It will never work. Just how do you propose to do it?!!! That is just more of that Buck Rogers science fiction stuff!

Well, at the beginning of all of the impossible jobs, statements similar to these were made. Have you ever heard the old saying that refers to eating crow? Well, a lot of crows have perished during man's process. The impossible jobs mentioned above were not impossible; they developed quite nicely and provide for primary portions of the economic systems that feed the world.

Just how do you propose? Provides the basis of this chapter. The writer thinks it is one of the best parts of story, because part of the story is a science fiction tale that permits kids from five to eighty-five to use their imagination to help design the project. For the engineers who actually get to work on the designs, if and when they happen it should be a real good job. The how gets into a few of what the writer calls the nuts and bolts ideas of actual construction—things that range from congressional approval to raising dairy cattle in high-rise structures. Science fiction or reality in the making—only time will tell. One thing that is certain, the chance to think about such a concept does not come around very often, and for many of us, never.

The idea of man building in the sea is nothing new. One of the oldest cities in the world is actually built in and above the sea, Venice. It dates back more than 1000 years and provides basic factual proof that construction in the coastal sea is real and can be very successful and profitable. Stories tell us the Dutch have used dikes and levies to drain land that is below sea level for a long time, successfully. The idea that we are discussing in One To Grow On might be thought of as the new space age version of successful construction and expansion of the past.

Building in the coastal sea areas that are protected is one thing, but when subjected to the full forces of nature, it is quite another. This is true, but man now has many advantages. Now he has the ability to build strong, tough structures that can dare the wind and water to blow; in addition to this he has experience and technology for designs that cover everything, including earthquakes. In addition to the abilities that exist, those that he doesn't possess he can develop on short order or notice. When we look at a globe of the earth and take a look at the coastal sea depths, then reference to the ideas we are discussing, we can form all kinds of additional ideas related to this type of construction. One point of view wants to tell us the more shallow the water, the easier and less expensive the construction will be—but on the other hand, nothing tells us man will not be permitted to develop the technology that will let him place the structures in water that is 1000 feet deep or deeper, after the methods advance and evolve.

If we look at or consider a map that shows the coastal sea areas of the U.S. and the water depth, we can form a few more ideas. In some areas the sea floor gently slopes for hundreds of miles before it drops off into the abyss of the ocean depths. This area is usually referred to as the continental shelf. The width of this shelf varies from hundreds of miles to almost nothing depending on location. The east coast of the U.S. shows a considerable amount of shelf area, in total, that ranges in widths of a few miles to hundreds of miles. The Gulf Coast shows a large area of fairly shallow water from Florida all the way over to Texas. The west coast is different; it doesn't show much of a shelf or shallow water. That does not mean this type of construction could not be used in or on the area that is available.

This far into the story we have most of the basics of the idea. It is a funny thing, looking back and then looking toward the future. The indications tell us, and some of the words also, that almost all of man's creative and inventive processes begin, simply, with an idea. The engineering process of the old days was, for the most part, all trial and error; in many cases it still is. Engineering has made miraculous and amazing advancements through the years—man's future, his successes and failures, appear as though they will be determined by the designers of the inventions and processes. We could say engineering does or will hold most of the keys that will take man into and through the future. Engineering is important; vital for mankind. Right and proper, it helps protect and create life—when used the wrong way, it sometimes destroys lives. It covers and includes everything from the correct bolt strength grade to identifying sounds and images from distant space that can warn of meteors, to presenting multi-billion dollar proposals to business and government.

No engineer in his right mind wants to design a death trap, with the exceptions of military or police defense and protection. When we keep this in mind, with reference to the projects we are discussing, it must be remembered that history and nature have proven how deadly sea and coastal sea storms can be; that some are worse than others, and it is only a matter of time until the next one hits. This should be at the top of the list for design specifications of all these things, placed in and above the sea.

When we get into the nuts and bolts of construction and design, it doesn't take much study to determine that man now has the ability to build almost anything. Even structures and buildings that can withstand 300 mile per hour winds, with little or no damage. Concrete and metal make such designs possible. In general terms, the costs associated with concrete construction are too high to permit wide scale use for storm proof construction; but, if the processes for the massive amounts needed for the uses we are discussing ever develop, it is easy to imagine costs can be made reasonable as the methods evolve.

If man is going to give this type of construction (in the sea) at least a chance for thought and consideration, it would be interesting to know a little something about how to put it together. Most of us, at the first mention of such an idea, might have a reaction that almost immediately wants to tell us no, that such structures in all of that water could not work. The reason seems to come from our general understanding, perception, or visualization at the very first thought, mention, or talk of such an idea. So, for the writer, one of the first parts of figuring out how to go about things, was to start with understanding and perception.

As an example: if we think about the sea, the sea-floor, and the solid earth underneath all of that water, it can change things. If we can imagine standing on the seashore, in most areas we would be standing on loose sand. Now beneath that loose sand is wet sand mixed with mud and sediment that has collected for years, in some places maybe even millions of years. Below that mud and sediment is bedrock; it is the outer portions of the almost solid crust of the earth. Most of the time it is strong and tough; and it's down there, in some places only a few inches, some places a few feet, and in some places hundreds of feet; but it is down there.

If we stand on the beach and look out over the sea toward the horizon, any combination of thoughts can come to mind. To assist us with our understanding and perception, we need to think of some facts. There is a lot of water in the sea. The water contains all kinds of stuff. First, science tells us all of that water is a combination of oxygen and hydrogen; then there are minerals and salts of all types and concentrations. Within this liquid mixture are all kinds of life that range from the largest whales to the smallest fish, and probably a few things man doesn't even know about yet. Underneath all of the water and sea critters is a layer of sand and sediment that compares to the stuff on the beach. Under that is the same solid crust of the earth. Well, mostly solid; like anyplace else on earth it has some cracks, voids, pockets of water, gas, and oil; some lava flows, a number of volcanoes, etc. But according to science, it is mostly solid rock down there about fifty miles or so.

There is a lot of land and solid rock under all of that seawater. The estimates tell us there is about twice as much land beneath the water as there is above the water; that is definitely a lot of land. If we could imagine being able to see all of that land without the water and its contents, it can help us understand. If we were standing on the beach at the same place and look out toward the horizon, in some places we would see gently sloping plains that go for hundreds of miles. In some areas we would see giant mountain ranges with hills and valleys; some places would have volcanoes, geysers and rivers. All kind of stuff; and an interesting thing is, we would be looking downhill. All of that land is below us. This indirectly tells us that we are standing on top of the mountain, even when we are standing at the edge of the sea.

The parts of this land that concern us most, with reference to the expansion and construction method that we are discussing, are the plains and gently sloping hilltops between the shore line and the next big valley or gorge. In some places we call this area the continental shelf. The only trouble is, it is flooded or covered with water, and man can't drain it off to the vacuum of space, yet. This makes the water an obstacle to construction and expansion. History tells us, and proves to us, man always and continuously, for thousands of years, removes, gets around, goes over, deals with, or uses any and all obstructions that get in the way of his or God's plans. That is a fact that cannot be denied. A comparison for this example is the shallow water covering the coastal land, in some cases and certain points of view, is little more than an obstruction to man's progress and gains going into the future; something that he will simply have to deal with—that is if man should make the decision to use the methods of construction we are discussing. As far as the fish and sea critters disturbed by man's expansion into the sea, it is a very big water hole; we can be pretty sure they will find a place to go.

Security, perception, and understanding all go hand in hand. Each one affects the other. In this next example there is a comparison between dry land and the sea. In general, man can do most of the things that he needs to do on dry land. When man stands on solid ground it produces a certain sense or perception of security. When it comes to the sea, man is limited in the things that he can do—simply because he cannot walk on water. This fact alone, regardless of all of the other reasons, affects our thought processes and perceptions when we begin to consider anything that places us or our financial investments in, above or below the water of the sea.

Let's try another example that can demonstrate perception. Suppose a well established and expertly qualified engineering and design firm prepared the plans and specifications for the first major project to be constructed in the coastal seas of the U.S. Suppose further that the plans are ready to be opened for bids to begin construction, but a very large sum of private investment is still required. Continuing, we have a conservative, successful, well established investment firm that has a few billion dollars that needs to be invested. A firm led by some old-timers that have helped build major cities and projects on dry solid ground, all around the world. But this idea of raising a brand new city up and out of the sea just doesn't sound right. The perception, combined with security and understanding, plus the known risks and hazards, do not add up and follow past standard practice close enough. The question is, will the old boys invest and build or will the plans be put on the shelf?

Consider the idea of investment, in the process of designing and creating demand for a large, new city on dry and solid ground. In general terms, everyone makes money and pays taxes; from the companies that bankroll planning and engineering to the kid sweeping up at a construction site—very few lose. Now when we invest to build on solid ground, we don't have many restrictive thoughts about the construction that is below ground-level. We basically accept the fact that if engineering installs things sub-level, so be it. In our major cities we have all kinds of things going on a few feet, a few yards, and hundreds of feet underneath the perceived to be solid ground that we might stand on or drive across—when in fact there could be voids, people and structures just a short distance down. There can be train tunnels, utility tunnels, roadway tunnels, sub-level structures of buildings, walk ways, and secret stuff—all types of things that are kind of out of sight and out of mind. Even though there could be hundreds or even thousands of live people, just a few feet down.

The idea of building a city on a structure above water is thought of in a different way. It just doesn't sound right, yet! Now if we can imagine a fully functioning city, with a population of a few hundred thousand people that has been built above the coastal sea, occupied and in use, our perception can change again; maybe to one of amazement or excitement that it actually succeeded and exists. Along these same lines we could imagine that those who work and live in a city on such a structure, would become used to it—concerning themselves with daily life, rather than contemplating being above the sea. Except, maybe after work they could be thinking about taking the elevator down to the dock level, jump in the boat, and go for a spin before dinner.

It is easy to understand that in order for serious large scale interest, involvement, and investment to take place with designs of this nature, the approvals and recommendations of engineering firms with the proper qualifications and experience would be demanded. When that can happen, we can be pretty sure that man's future will take a turn for the better.

Now a short change of subject. At this point of the story, during one of the many rewrites and close to the final revision, an odd thing happened. One of the news reports aired a story about a large whale that was swimming in the Hudson River, over there by New York City. A story about a whale in the neighborhood is not the type of story that we usually hear out of N.Y. Such an event makes the writer smile. It kind of fits in with the parts of the book that relate with nature, spirit, faith, and old man time. One of God's largest creatures, swimming in the coastal sea waters, between the Statue of Liberty and one of the largest financial centers of the world. Like the kids used to say, it was pretty cool.

Back to the discussion of perception. If, or when, man successfully builds one of these massive structures in the coastal sea, and it is put into use and occupied, we could expect that man's understanding or perception of such an idea would change. Change in a way that would make it much easier to accept the farfetched idea of this type of construction, due to the actual success and physical proof.

We can easily understand that, for serious, big time, investment consideration of such ideas, man's basic perception and understanding will need to be altered in a number of areas or fields; in comparison with the established guidelines and methods of the present time period, along with those of the past. First, simply, is the thought of going into the water with any kind large scale construction. Next are the new methods of construction that do not exist for the quantities of material we are discussing, that would need to be developed. It is not too hard to understand that all of the basic technology and machinery, or building blocks, exist right now—that would permit man to expand and refine the processes to develop the massive quantities needed. Then we get into the areas of financial costs and time periods of such long-term projects. Such costs and time periods fall way outside of the normal or everyday planning guidelines most people concern themselves with. Just to be able to think about such ideas seriously requires fundamental changes in thought, with respect to existing standards, especially so with our professional and conservative business experts.

If we attempt to imagine or look far into the distant future, with this type of construction in mind, one point of view wants to tell us a possibility does exist for man to manufacture and construct areas above the sea that could compare in size to small states and countries. With this in mind, and referencing much of today's science, in a way, it is about 180 degrees the other direction. Much of the current science involves the atomic and sub-atomic structure of matter, the small stuff, so to speak, that has provided the means for many things. One of the most amazing is the ability to generate an image of the great expanse of the known universe—based on scientific fact. One of the many others deals with the structure of matter, and the ability to create composite materials that are stronger than steel. Consider this: science tells us the solid rock of the earth is about fifty percent oxygen. The rest is a combination of various minerals and elements. Then we add in the ability of science to create light weight carbon structures and alloys that have great strength. It demonstrates that the door is wide open for science to get involved; to create new and unknown methods and processes for the type of structures that we are discussing. This, combined with the proof of man's developments to this point in time, wants to tell us man will find or discover better and less costly ways to produce or manufacture the structures we are discussing. History indicates once man gets processes started, at some point in time his production costs are reduced by large percentages, after the development and advancement of processes, methods, demands, etc.

Some of the technology that we know of is, in one word—amazing. Yet, some of the science and technology that exists and we are not aware of could knock our socks off . . . almost. On top of that there is always the newest stuff that is locked up and known only by a few—that even when made public, it seems as though it is beyond belief. Again, as we combine all of the bits and pieces of science, fiction, and the proof of the past, with some thought and peeks into the future, it almost gives us proof that man will be building some real big projects in the coastal seas. But a storyteller is a storyteller; don't bet the farm on the idea, yet.

The story touches on many subjects and topics that, in some ways, have something to do with presenting this idea of expansion construction. This part about perception developed some place between the third to fifth revision. (My sidekick says it was probably during the fifth . . . fifth of liquid spirits that is.) I kept asking myself what it would take to make the idea believable, then this bit about perception just kind of developed. It's a little like trying to believe in 1930 that within a few years man would have a jet that could break the sound barrier; or in 1950 trying to believe one of the hot dog fly boys would be walking on the moon within a few years. Yeah, it's kind of like that. As far as the relationship between nuts and bolts and perception goes. Well, we might think of perception as a fundamental nut and bolt of our thought process.

* * *

One more example of perception and how it kind of tricks us. In our normal day to day lives our concerns do not permit much time for our thoughts to wander into and ponder areas that don't serve a useful purpose for getting through the day or the short-term, near future. But this example can tell us something. Consider this: most of the time we do not consider the fact that science tells us we are located on the side of this huge rock called Earth. A rock that is suspended in space, revolving at a speed of someplace in the neighborhood of 1000 miles per hour. Nor do we give much thought that at the core of this rock is a continuous chain reaction of stuff that produces a liquid of molten minerals and materials that are estimated to be in the area of 10,000 degrees F. Nor do we give much thought toward the facts that tell us that there is hot molten lava under our feet and below the fifty miles (or less) of the outer crust of this big old rock.

We can suppose that everyone considers these facts in a different way. Some of us will simply say to ourselves, ok, so what. Some of us have never really thought about the idea and might even find some amazement in the thought; and others who rely on faith might say something like, yeah, wow . . . we know—ain't it great!

I think this provides a good example of how perception and understanding can change.

* * *

One of the largest nut and bolt portions of the ideas we are discussing boils down to a bottom line of massive quantities of rock needed to make concrete, steel, alloys and other composite materials. Another part of that same bottom line wants to say to those whose want to pursue this idea, the crust of the earth is about fifty miles thick and it should contain all of the rocks necessary for man to build projects from now until the end of time. All man needs to do is figure out a way to make it happen. Even with today's standards, technology, and equipment, some of the old underground miners will truthfully tell you that they can mine tunnels all the way around the earth—if you can supply the manpower and equipment. Some of the steel makers will tell you something similar that says, they can make all the steel you can use, if you can provide the manpower, equipment, and materials—materials that miners produce. In a roundabout way, this is another one of the indicators that tells us there is no excuse for not developing the ideas being discussed. Still, it is all much easier said than done.

The continuous amounts of materials needed for such construction is way beyond anything man has produced so far; and by current standards pretty close to impossible. But not beyond the science fiction that brought you everything else that exists today. To give us a reference for the size and quantity of the support structures that rise from the sea floor, with the ideas we are discussing, try to visualize two structures made of concrete and steel that compare in size with two Hoover Dams. Imagine them in the coastal sea a couple of miles off shore; they could be in about 300 feet of water. Imagine the structures are about 500 feet apart and parallel to each other; if you can, imagine a thick concrete slab maybe fifty or a hundred feet thick spanning and covering the tops of the structures—strong enough to support almost anything. This provides the basic idea of the methods of construction that we are discussing.

Current standards of production, construction and finance almost immediately want to tell us—no, can't be done. Well, we know for a fact that man and time change most of those "can't be done" jobs, into successful "can-do" jobs. The writer stumbled onto an idea recently that provides some stuff to think about. This has to do with what the boys used to call, getting twice the bang for the buck, killing two birds with one stone, dual use or purpose, and getting twice as much for your money.

Reports over the years have told us something like fifty percent of the total populations live and make their homes along the coastal sea areas. For the story, we can suppose this to be fairly accurate. When population densities become overcrowded, all types of things are affected—even the numbers and ratios that define overcrowding. One of the areas most people are familiar with concerns the highway and roadway system. If man can combine the need of new expanded highway systems with the need for large quantities of rock needed to make concrete, he might be able to engineer one of these two-for-one deals, that has a chance of being very profitable.

We are getting into an area where at least two demands on man, in the future, might have a chance of being met by placing large portions of future highway systems into sub-level or underground tunnel systems. Especially the long distance portions of the interstate highway system that might restrict access and exits to limited numbers of interchanges. The possibilities and the required study of such an idea would fill a few books. But right away, we can find that possibilities do exist for designs that can produce rocks for concrete and at the same time produce something else that could have value—and in this case, secondary values could be very large.

For this portion of the story, One To Grow On, we are basically discussing ideas that produce materials for concrete and steel. But let's take a brief look at this idea of placing some of the highway systems within tunnel systems, and mention a few highlights and thoughts. One of the first thoughts that comes to mind is that if the tunnel type highway systems are planned as toll roads, logic tells us at some very far distant date or period in time the toll fees are supposed to pay for the construction costs. This point of view indicates that the material or rock that was removed for construction purposes, or making concrete, will eventually be paid for by the secondary use that collects toll fees. But the time periods are not realistic, with regards to current standards. Arbitrarily, we might pick time periods that range from 100 to 300 years, as pay off periods for a four-tunnel highway system that runs from Boston to Miami. (These numbers are not accurate and are for example purpose and thought only.) When we consider installing highway systems in tunnels, a number of other benefits need, at least, to be mentioned. One of the first is that such systems are protected from the weather—no snow to plow or to cause accidents. No ice, no rain, and no sun to boil the oil out of the asphalt. All of these have huge monetary and time values associated with them. If we consider the latest developments in the electronics and computer fields, it isn't hard to imagine that long distance, limited access systems could incorporate automated guidance, speed and spacing controls—virtually eliminating accidents and some highway deaths. If we consider land costs, future appreciated value and future need, one point of view indicates that by using tunnel systems man might actually create value gains, especially in or around large metropolitan areas. Remember, this is part science fiction and these are simply ideas that would require study; but such ideas demonstrate dual purpose or multiple purpose types of engineering that would be necessary, and profitable.

Using a certain point of view, we can sit back and make a general statement that says the crust of the earth probably has all of the stuff that man will need to pull this off, if and when he decides to. Such a statement, for the most part, would be true if he can develop the processes and machinery necessary to make it happen. This particular if, is a real big if. While our old miners can tell us stories about mining tunnels around the earth, they would also tell us that they would produce only a small percentage of the material needed for large-scale use. But man might think of starting development with such methods, because of the potential for creating additional value with multiple use engineering. Then, as production demands increase and the bugs get worked out, we could expect man would start removing a few mountains from the surface of the dry land in addition to the mountain ranges that rise from the sea floor. It is easy for us to imagine that it will take quite a few years of development in order to produce the high capacity machinery and equipment that can reliably function at depths, pressures and temperatures found three to five miles below the surface of the sea or the surface of the land.

Let's suppose that, someplace in the future, man decides to take a shot at this. His studies show that he can drive a highway tunnel system from the Boston area to Miami, with interchanges at the major cities that will do the job and make money in the process. In addition to the highway system, the rock removed will provide a few cubic miles of rock to make concrete—enough to start a couple of the coastal sea construction projects. It could even be that his studies showed it would be less expensive, in the long run, to drive through tunnels, as opposed to the conventional highway systems. His tunnels might go real deep, 1000 to 3500 feet below the surface. By going to such depth he can place the tunnels almost any place, without need to worry about surface disturbance. It means he can go below the towns, cities, rivers, mountains and swamps—and not one bridge, except at or on access ramps. It could be that in order to get the right amount of production necessary to make the structures in the sea, the tunnel system would need to be divided into twelve, twenty-four, or more sections, that were worked from both ends or directions. If we try to imagine the coastal sea projects, the largest percentage of the off shore transportation process during construction would be by some sort of floating vessel: ships, barges, tankers, etc.

When we get one to three thousand feet below the surface, we get into the solid crust of the earth—stuff that makes strong concrete and good tunnels. But going to such depths would require long access ramps, or short ones with a lot of curves, in the process of building highway tunnels as such depths. As an example, if the tunnel was 3000 feet below the surface of the ground, at a place where man wanted to have an entrance and or exit ramp, if built in a straight line from the tunnel to the surface, it would need to be something like nine miles long to keep the grade from being too steep. But, by adding a few curves, the length could be shortened and the opening or openings could be placed in almost any direction from the tunnel. Again, as a toll road, it is supposed to pay for itself at some point in time; and if designed to function as a true high speed expressway it would be desirable to use for many, making it less painful to pay the tolls. The important thing, for our original story, is that such a tunneling project could provide large quantities of material for starting a coastal sea project.

The idea of going into the ground for survival and economic purposes is nothing new. The old caveman started it all when he dragged his woman into the cave and started making a family. Sometime after that history tells us the Roman Empire started some of the first mining projects that were designed for water tunnels. Here in the States, the coal miners get most of the credit for starting the development of production and safety standards for just about anything that goes on in the ground beneath our feet (and most of it was learned the hard way). The old gold prospectors out West perfected hard rock mining, different from mining coal. Then you have your sand hog miners over in the Northeast; they drive all the subway and utility tunnels. All of these men and boys have produced some amazing stuff in the past hundred years, a lot of good value in the process. Most of the ideas in this book are not really new, they are just a little different from the current standards and methods of doing things. Like one of the old boys said, "the next big thang."

* * *

Highway systems in tunnels is one idea that could

help produce and pay for part of the rock needed for the project. If we dig around a little (pun intended), we might find a few other ideas for making a profit with tunnel systems. Here is another example. Every year, for hundreds of years, flooding has been a problem that destroys lives and property. In recent years, the costs related to the destruction runs into the billions of dollars. When major river waterways begin to back up, the problems increase upstream. Over the years, dams, locks, dikes and levies have been constructed to control the worst of the problems; but larger amounts of destruction still occur. Common understanding wants to indicate that if man could prevent such destruction, the cost amounts could be invested into something constructive and more profitable.

This says anytime man can do something to prevent or reduce major flooding, he reduces destruction and has a chance to save lives. Going further, it indirectly tells us if man could find a way to actually alter and reduce the flood plain areas associated with the major waterways in the country, he could do something good and probably make money in the process.

Let's tie this in with the tunneling and rock production systems we are discussing. If man of the future starts driving tunnel systems, he has another possibility for a secondary use. It would involve designing tunnel systems deep underground in the earth, that could act as overflow waterways; they would connect with the major waterways at the places called choke points. Tunnel systems such as these would need to be at great depth and a downward angle of a few degrees, to some out flow point at a lower elevation. They might empty into the sea in some areas, might dump into existing lakes and reservoirs, or into new ones that would need to be constructed. We could imagine that once such systems were installed and proven, flood plain areas could be reduced in size. This could open large new areas to development; destruction costs associated with flooding could be reduced; insurance costs could be reduced; and maybe a few lives saved in the process. It's another one of those two-for-one deals.

When we peek into the future, we can find indications that hint of future problems for man that have a chance to relate to these tunneling, or mining systems and projects. The new science of the past fifty years or so, provides us with all kinds of things to think about. One of these is the information that has been produced about the sun. The information tells about the size, power, sun spots, solar flares, solar radiation solar wind, particle streams and barriers, magnetic fields, magnetic storms, etc. It is a long list that includes a lot of stuff. Most of us have learned about some of the power of the sun the hard way, having experienced a case of sunburn at some time in our lives. It is a very powerful ball of energy.

Out of all of the data and reports one area stands out: it has to do with electro-magnetic field fluctuations here on earth. Very powerful fluctuations that are reported as having caused electrical power and electrical communications interruptions and outages. This single threat or problem, especially if prolonged, could produce some very nasty stuff. The reports, combinations of science fact and science theory, tell us that events within the mass of energy we call the sun, in recent years, have induced magnetic field changes on earth that were strong enough to force electrical generation and power systems into emergency shutdown, along with communications systems. When man attempts to shield these systems from the power of the sun there is only so much he can do. Most of us have some personal experience and ideas that help us understand the problems that occur when the power, telephones, computers and radios no longer function. When such events shutdown major cities, regions, etc., it gets real serious, real fast.

The tunnel or mining systems that we are discussing with the story have a chance of providing good protection against some of the destructive power that the sun throws, blows, induces, radiates and sends our way; especially if it increases and starts causing serious long-term problems. When we get inside the earth and start putting hundreds and thousands of feet of solid rock between us and the powers of the sun it shields man from many of the destructive forces. Science tells us about a few sub-atomic particles that pass through solid rock and travel deep into the earth, but not many. This shielding effect also applies to electro-magnetic fields, to some degree; and that would require professional study to determine the effect it could have toward protection of power generation and distribution equipment, including the high voltage transmission circuits, along with protection that might be offered or made available for communications systems. The bottom line says man might be permitted to protect his power systems if he installs them inside the tunnel system that might need to be developed. On one hand, it is not that big of a deal; it's kind of guaranteed work and economic supply to carry man through the future, because he has the science and the tools to get started. On the other hand, if he should fail to use all of his science and tools to do the things needed to provide protection we can use our imagination for that also. Another interesting thing to note: if, and this is another big if, such tunnel structures should ever be required to provide this sort of protection from the forces produced by the sun, it would probably make the use more important than any other use man could invent or design. This indicates the production of rock for making concrete would then become a secondary benefit of building the tunnel systems, like an added bonus, making it another one of those two-for-one details.

When it comes to rocks for making concrete, man has all kinds of sources. If we think of the sea floor, science tells us that some of the mountain ranges that rise up are taller in height than the mountains man has up here on the surface; and that undersea volcanoes are constantly building more. This tells us cubic miles of rock are waiting for man to figure out a way to make a little economic supply with them. Stories tell us mining does take place in the sea, but the machinery to move the quantities of material we are discussing we should think of as waiting to be developed at some point in time in the future.

When it comes to building machinery, man in 2007 has the ability to build some pretty fancy stuff! Man has the experience and capability to fly remotely controlled equipment to Mars and beyond, and watch it on a television screen. If that is possible, it almost provides the proof that he can build the large machinery that will be needed to work the sea floor. We could expect such machines to develop pretty quickly, once the first successful demonstration project gets started.

Now if we go way out there, with a little science fiction and some Buck Rogers stuff, we can find a few more things to ponder in man's future. Man has drilled, dug, mined, and poked holes into the earth for a long time. It was only 100 years back when Mr. Rockefeller started the oil industry. Well, to make a long story short, some pretty fancy drilling machines have developed over the past 100 years! As I recall, from one of the stories, one of the deepest holes into the earth is something like eight miles deep—that is a deep hole! In comparison, imagine the types of advances in machinery that the boys will make in the next 100 years. Just for the purpose of speculation, let's suppose that man will have machines that can drill twice as deep, say sixteen miles. In some places a hole drilled that deep into the earth will be able to tap into hot flowing magma that is under pressure. Magma, in simple terms, is hot liquid rock. For the expansion and construction methods that we are discussing, the material that is used and needed is rock; that is the moneymaker. If the new drills work and the boys manage to get hot flowing magma to flow to the desired location, we could tend to think that they might have it made. Another way to think of it is the profit margins might have a chance to add up and spread out in ways that could compare with the oil industry; except, much greater. It is definitely a long shot, but imagine a few of the possibilities!

We can stretch this bit of fiction out a little further. Man has all kinds of stuff in holes in the earth, all around the world, such as sensors and gizmos to measure temperatures, seismic events, pressure, etc. All of his science gained with such devices over the years has produced some amazing fact and theories about the things that go on inside the earth and the physical structure. If we combine all of the science with the nature of man, plus his advancements with machinery, then add the problems that seismic events or earthquakes, and volcanic eruptions cause, the indicators want to tell us future man might have a chance at the development of methods to control such destructive forces as earthquakes and volcanoes. It depends on how much he can get out of the new drilling machines he might build in the future. Imagine the possibilities of being able to drill into the pressurized magma chamber of a volcano to control and relieve the pressure, and prevent an eruption! Suppose man's future advancements include the ability to locate the pressure points on the tectonic plates; that might be drilled to relieve the pressure and prevent earthquakes, wouldn't that be something? Let's hope that such advances can and do happen; and that the boys don't poke a hole in the wrong spot, that might cause the earth to take off into space the way a balloon takes off when punctured. The ability to tap and control magma flow—that would be something!

If man ever decides to develop the construction and seawalls we are discussing, there is always a chance people in the future will change their way of thinking. Could be, future man will decide to go into the Rocky Mountain West and just start loading a few of those 12,000 footers, that have easy access into trains and ships them to the job site. The cubic miles of material would add up quickly that way.

Man, in the future, should have quite a few options for developing materials and turning rocks into concrete and metal; it looks like we covered most of them. But the idea of developing high quantity tunneling methods seems to have very special importance—especially with regard to shielding from solar activity and magnetic storms. Importance that indicates future man might really need them.

* * *

Primary Supports

The primary support structures used with these methods of construction (and the seawalls) would be very important. We briefly discussed the massive sizes and how they would be used. If we try hard, we could almost visualize such structures rising up from the sea floor. Structures such as these must be properly engineered and constructed. Any errors that might occur must be on the side of strength and safety!

Structures such as these have one primary purpose, to permanently support and withstand all loads and forces that might be applied to them; by doing this all loads and forces are transferred into the crust of the earth or the ground below them. In order to transfer all of the applied forces and loads without any problems, they must be made of the right stuff, be of the right size and shape, and be installed properly. These support and deck structures should be thought of and engineered as extensions of the earth, that would support almost any load or force applied to them.

The "right stuff" is not a technical term that is commonly used in any field that the writer is aware of. That makes it hard to list in the specifications for design. As a definition for our fictional story, we could say that it is the result of all of the known components and processes that combine with all of the things and forces that are not known or realized that produce success. These portions of success that cannot be measured or specified reach into areas that include things like knowledge, determination, faith ethics, God; even the interaction between sub-atomic particles that hold stuff together—including those that man has not discovered yet.

The primary component of such structures that

transfers the weight/load and any applied force into the ground or crust of the earth, is the reinforced concrete mixture. That almost makes it the most important part of the structure. But the proper engineering specifications that produce the concrete, along with all of the related processes, provide the primary foundation for success. Concrete, truly, is amazing stuff. It is made from mixtures of rocks, sand, dirt, water, air and sometimes made stronger or reinforced with steel or other alloyed metals. Included in the rock and dirt mixtures is a product called cement. This is a scientifically engineered mixture of various types of rocks and dirt that help form the bonding qualities; said another way, it makes it all stick together. In a way, engineering concrete is a science of its own that calls for specific mixtures, to serve a wide variety of applications. All of the years of experience and new science have provided man with the knowledge of molecular structure and sub-atomic interaction that produce the bonding, or stick together qualities of the various mixtures. The bottom line says man has the ability to make good strong concrete that can form structures to do the job.

The physical shape of the structures will, or would, affect the way that loads and forces are transferred into the earth or ground. It is easy to imagine that square-shaped structures of massive size would be the strongest or best way to support the applied loads. But if the shape can be altered and connected to the complete structure, in a manner that supports all of the applied loads and forces—that only requires a fraction of the amount of material needed for a solid square structure—the result is a savings of time and expense, that equals profit and economic supply.

Consider the pyramid shape. First of all, it would be

interesting to know how the ancient rulers of the past came up with the idea to build structures in such shapes. One of the stories said it had to do with angles and the way spirits return to heaven. The influence to build the first pyramids must have been a very powerful influence.

Now getting back to our project. If you have a one dollar federal reserve note handy, there is an image of a pyramid on the back side. If we can imagine a pyramid-shaped structure, maybe 700 feet tall, without the top 100 feet or so, that would make the point; and imagine the base as being maybe 700 feet corner to corner, and having such a structure placed in a coastal sea area in maybe 300 feet of water. If we can imagine all of this so far, we should be able to imagine the ocean waves rolling up the sides and the corners of the pyramid almost cutting the waves as they begin to push against the structure; then the waves roll back down the side and the process repeats over and over through time. If we try hard we can almost imagine how the weight and forces of the moving water transfers through the side of the structure into the base and into the ground. If we can permit our imaginations to wander a little further, try to imagine the worst storm blowing against such a structure, made of concrete and steel. One point of view says the structure will not be affected, and it is strong, tough, and there to stay; the tapered sides of such a structure would permit the water to move around it more easily, as opposed to a solid square mass. It's almost as if the moving water of the sea and the pyramid shape were made for each other; it fits.

Supposing and imagining are one thing—but engineering load forces, mass density, weight distribution angles, along with all of the other technical factors, will determine if a structure of any shape or size can do the job. From most of the indications we have every reason to believe that man can find a way to make the idea succeed, with any shape needed or desired, if he chooses to follow through.

Each square mile of flat surface area of these structures would require a number of primary supports. The number required would need to be determined by expert engineers. For the story and as an example of strength, mass or size, along with a few facts, we can use Hoover Dam; it is located in the West on the Colorado River. It provides some good stuff that serves as a tool for thought. Any large concrete structure could be used for comparison; I chose this one because the facts were handy and I have had a chance to see it.

First, the government built the dam. The published numbers tell us the dam was approved for construction in 1928 and completed in 1936; the cost was about 175 million dollars; it contains about 4.5 million cubic yards of concrete; it is about 1,240 feet long and 725 feet tall; the entire project cost about 385 million dollars.

It is a good, strong structure; it is very important. It provides water for many uses; irrigation water for some of the vegetables many of us eat is made possible by this dam. The structure was built in place between the canyon walls; it took about ten years to draw up the plans and get it built. It started with an idea.

Between 1936 and 2007 a lot of things have changed except the structure itself. The structure is positive proof of the success of the engineering, construction, strength, and function of the idea that it all started with.

In 1936, 385 million dollars was a lot of money. When we look back at all that has been produced and provided for with this structure, we can almost say it would have been a bargain at any price. Now consider the new structures and methods that the story discusses, along with the high costs involved. Do you think they might have a chance to pay off in a comparable way—that in fifty or seventy-five years, it could be said that they would have been a bargain at any price?

The structures needed for the project require a way to manufacture them, and not one every ten years! Let's suppose the studies prove the methods being discussed could work, if man can produce these primary support structures quick enough; and at reduced costs, when compared to today's conventional standards. All of our history and indicators tell us when man needs to produce faster and at reduced cost, he starts looking at mass production methods, of some sort. The writer thought this over for a while and came up with an example for an idea.

The production method might go something like this: all of the reinforcing steel needed for such structures would be welded together at a shipyard; then a steel or alloy, waterproof/leakproof skin, hull, or covering would wrap around the metal reinforcement structure. The skin, hull, or cover would need to be designed to serve as a concrete form; this watertight structure, whatever shape it might be, would be a vessel. Such a vessel could then be towed to the project or desired location; once in place and prepared, concrete pumping operations could begin to fill the concrete form or vessel with concrete and sink it into position. That is the bare bones basic process. Every time it gets multiplied by a factor of 10, 15, 30, things start to add up pretty quick.

Well it's not that simple. The proven facts tell us, yes, man can make the steel and metal for the reinforcement structure and build the vessels; and yes, man can make the cement and mine the rock to produce concrete, and he has the pumps to start pumping it. But the indicators tell us size, quantities, production times, and expense are too extreme compared with the current standards of production, manufacturing, and finance—at this point in time. So how long will it take to develop the capabilities to get started? Like the old boy said, that is the 64,000 dollar question. Here is one way to think of it. History proves beyond any doubt that development and change has, can, and does happen pretty quick, especially when sums of multiples of billions of dollars are involved.

Now let's slow this down a little and take a look at this vessel idea. Sometimes if we can visualize things, it can help our understanding. Some of us have seen the large ocean-going ships, and some of us have only seen pictures of them. Some of the big ships are the length of three football fields, end to end. Try to picture one of the large ships, such as an aircraft carrier, offshore in the coastal sea. This next part is stretching it, but stick with me. Imagine if man was to pump that aircraft carrier full of concrete. In simple terms, it would go straight to the bottom and nothing would move it, except for a catastrophic event or a very large explosive device. This simple illustration provides the basic idea of using floating vessels as concrete forms in the process of building the primary support structures we are discussing. In addition, from all indications, when the need for large scale seawall construction starts, the same type of process should be able to help man with that problem. Something else to take note of with this example: the Hoover Dam compares in size and cubic capacity to that of the aircraft carrier used in the example.

Suppose man had a project that called for 1500 primary support structures that compare in size to the Hoover Dam, and he intends to use this floating vessel/form to do the job. It's easy to imagine that it would take a while to get them built. Imagine a little further, that the next couple of projects require say 10,000 of these primary supports. This provides us with a quick and basic example of the types of production capacities that might be required. When we consider future man might be demanded to build seawalls with these same types of structures, we can begin to realize the challenges, and need to develop such processes and abilities far in advance of the time period when they will be needed.

It's pretty easy to imagine the primary support structures hold most of the keys of success, for any type of large scale construction in the coastal sea waters. As far as building cities on top of flat deck structures, let's put it this way. Man shouldn't have any problems designing and building the cities. He has plenty of experience, knowledge, tools and money to work with, now. Man has been pouring concrete into flat decks for a long time; he shouldn't have much trouble with those either.

Indicators want to show us there are not many options for placing large concrete structures in the sea, other than with a prefabricated form or vessel; comparable to the ones we discussed. When the time comes that man will be demanded to prepare for the construction of seawalls, these floating forms containing the reinforcing metal, should do the trick. But if we make a very rough estimate that says man might need something like 100,000 miles of seawall to protect the world's coastal mainlands we can use the number for a reference that can help us judge the size and time periods involved with such a process.

Man has done amazing things with his mass production processing using automation that incorporates computerized controls over the years. When automobile manufacturing started, they were built by hand—one at a time. The first ones took years to build. Advances were made, then they could be built in a matter of months. Advances continued to the point that they could be built in a matter of hours. Now man has processes that can push brand new cars off of the assembly line at the rate of one each minute—one minute! I still find that amazing!

When the time comes that the huge concrete forms must be produced, they would need to evolve or advance in a similar manner. At some future point in time automated processes might be required to produce these types of structures in a matter of weeks, then days; as opposed to years and months.

Well, that is one point of view that covers the hard part of how this type of construction might develop. In the most basic terms, it is simply a matter of building the concrete forms, filling them up with concrete, and sinking them into place.

* * *

Financing

It is a funny thing. When we apply a little of that "twisted" logic, everything that has ever been built, manufactured, or produced was financed, one way or the other. Sometimes at great profit, and sometimes at great loss. Statements such as this don't supply anything that is of much use; but this does indicate something worth making a note of, because it is true. So, if we take it a little further, if man decides to use such methods of construction, or, when he must develop these methods of construction, the means to provide financing will be developed.

Even so, it does not help us much in an attempt to form ideas for financial planning.

How would man pay for all of this stuff? That is a good question to have the correct answer for. This writer can provide a few ideas, examples, comparisons, and that's about it. The right answers, details and specifics must be supplied by the experts.

I believe the term Uncle Sam started with the military back around 1800. The term has expanded over the years and sometimes it is used to describe the entire U.S. government. My use of the term is directed toward the protective and benevolent portions of the government. There are different terms for some of the other portions. Old Sam and his spirit have been around for a long time. He has helped finance everything from backyard gardens to the purchase of the western half of the U.S. Old Sam knows a few things about real estate and business.

One of the best ways to explain some of the ideas is by using examples from history and past practice that were successful. We can briefly study just about any major development of the past that involved government financing and draw a few general conclusions. Even with this type of brief study, general conclusions and simplistic ideas can help in forming real and substantial ideas.

If we go way back in time to the early years of the development of the U.S., things were much different. But success is success; it is a quality or factor that is timeless. It is interesting to note the country was built and still functions on the success that was created many years ago.

Way back in the beginning private companies, individuals, government divisions and leaders would get the idea for some new plan, process, invention, etc., then develop models, prove methods, and provide details for plans. Then, with some of the larger projects, a little arm twisting and back slapping might get a representative or senator involved, to see if he could help in some way. Basically, this was the procedure that helped build the country. Well, things have changed quite a bit since then; but once we get past all of the committees, agencies, go-betweens, etc., the basic process is still about the same. The old Uncle Sam portion is kind of in the background. Included with all of the good he provides, we could think of him as owning a small part of just about everything. He is an investor. We can be pretty sure if something is going to make money, Old Sam will get a piece of it one way or the other. If this type of construction, in the coastal seas, ever develops, we could expect it to follow along within the proven and successful guidelines of the past.

In an attempt to get to the bottom line, ideas will develop and be studied and when they have a chance at making money and/or providing security, this same process will repeat a number of times until production and acquisitions of greater and greater proportions take place.

One of the general conclusions tells us, all large scale financing consists of combinations of investments, bonds/loans, and taxes; said another way, all of the big jobs are funded by these sources. But the one factor that has the most power or effect is time.

We have a few lessons of history that can prove the success of large scale expansion projects of the past, that demonstrate how the combinations of government and private investment built this country, and a large part of the world. By discussing a few of these, maybe we can gain a little more understanding about how things work for the good.

One of my favorite examples is the purchase of the land and area known as the state of Alaska. History tells us the whole state was purchased for 7.2 million dollars; that was about two cents per acre. The year was 1867, only 140 years ago! One man gets most of she credit for making the purchase happen, Mr. Seward. Those who opposed the purchase called it a variety of names such as Seward's Folly, with references to things such as ice-boxes, penguins, Eskimos, igloos, etc. The bottom line is that it was an excellent investment that paid off. How well did it pay off? One point of view wants to say that the state is priceless; it has so much value that it cannot be measured. The value of those 7.2 million dollars, in today's dollars, could be in the trillions of dollars range; it's hard to say. There is no mistake. Mr. Seward was working for old Uncle Sam! That was one of his very best investments, ever.

How much value do you suppose has been created by it over the years? Better yet, how much more value will be created in future years? Suppose investment into coastal sea expansion and construction might produce profit in a comparable way. That would be an investment old Uncle Sam could be proud of. Seward's Folly . . . not hardly.

Do you think that was a good one? Well, old Sam's boys pulled off one that was even better back around the year 1803. This little deal was no pig in the poke either. It was called the Louisiana Purchase. The price tag assigned to this one was only fifteen million dollars. This investment purchased the land that makes up part of fifteen states, sixty-four years before the Alaska purchase—no wonder they called it Seward's Folly. It cost half as much as fifteen states. How much wealth and value do you suppose this investment has produced in the past 200 years? How many million people do you suppose call these two investment purchases home now? If new cities start popping up in the coastal seas, how many might be calling them home in 200 years?

Let's look at one of Sam's recent investments, the interstate highway system. History tells us construction started in 1956 and cost about 100 billion dollars over a forty year period of time. Billions of dollars to build roads across the land he paid fifteen million for about 175 years ago. Road construction alone was worth more than the purchase price of the land. Consider this: our story is discussing manufacturing the land and the cities, and we haven't really discussed the roads or transportation systems that could be built and used. How much value do you suppose those roads might produce? More of that logic with a twist, but worth a little comparison thought.

Consider the interstate highway system. Most of us take it for granted and cannot imagine not having it. In 1950 it was a true investment in the future. President Eisenhower gets a large part of the credit for making the construction happen. Today, most of us could not exist or survive without it. It is very important. We must wonder if any of the people who designed the system could foresee just how vital and important it would turn out to be. Take a while and study the combined and associated value related to the highway system; it is amazing.

These three examples compare with many of man's successful developments through time. But there is an unknown portion to all of this that wants to tell us these types of actions are driven by need and something else—that we can't quite put our fingers on. People from all walks of life use terms to describe such things; terms that include fate, destiny, faith, God's will, and thought that tells us some things are just *meant to be*. In these examples, and many others through history some of the indications tell us they could not have happened any other way. The amazing thing is, the financial portions were taken care of. As if to say, the events or investments were going to happen one way or the other, and they worked. When we think about the future in this same light, we can look at this story, with the huge quantities and costs, then apply the *meant-to-be* factor, predestination, fate and/or God's will to form a general conclusion. One that can tell us, if such things are in man's future, we can be pretty sure that the means to finance them will work out (almost) just the way they should.

If we use a certain type of viewpoint to briefly study these three examples of man's successful advance through time, we can connect them with a common event line or a common thread. Common in that these events or investments were the starting point for primary expansion, construction, and all around good stuff—in man's timeline of history. One of the viewpoints can tell us when the decision to sign the papers that made these events happen or begin, was the moment in time that many good and great changes started laying the groundwork for the good that we have in 2007. Another common point is each of these events had a price tag attached to them. That makes them economic in origin. That is part of the reason they were selected as examples. Another reason is because war and the hell of war are not highlighted in these events. It is possible to view them as peacetime expansion projects that were very successful. Yet on the other hand, wars of the past made these events possible; and in some ways were for protection and defense of future wars and the hell that they might produce.

With this in mind, let us consider the timeline of history. The timeline of history contains everything from the past; it would be possible to get absorbed by it and lost within it if we were to permit that to happen. So, we will get what we need and get out. Consider the three events we have discussed. Let's place them on their own separate and parallel timeline. This isolated and parallel timeline starts at about 1800; it only has three events. Let's extend it over to 2007. Now we have many common threads that connect with these events, but if we are going to make any sense with this we need to be selective and isolate a few.

The common threads can help us form a general conclusion that says they were primary event points when the seeds of change were planted; they can be viewed as investments that had a price tag; at the core they were peaceful in nature; they provided and promoted great amounts of peacetime expansion; they generated great prosperity that continues to grow, producing good effects all around the world.

In 1956, the beginning of the interstate highway transportation system marked the beginning of an automotive industry and transportation system, it provides the core of the world's economic system that exists today. A price tag attached to it says 100 billion dollars. If we try to visualize these two timelines, we find that we could add quite a few events that match the common threads of these three. They were all important. But we need to be selective so that we can get to the present time and give some thought to events of the future that could match the threads of good things produced.

If we think of our timelines between 1956 and today, many events occurred that mark the point in time when the seeds of change were planted. If we look for the next one after 1956 that has the largest price tag, it almost has to be space flight, with everything that is related to it. One quick conclusion is the prices keep getting larger.

That indirectly tells us the next big event will need to cost and produce more than space flight has so far, and that will be very expensive. Expenses that might compare with raising cities up out of the sea and seawalls.

After we study all of this stuff, combine it, and shake it out, man, in 2007, is at the edge of another big event or major discovery—something special and something big. Now, if we bring old Uncle Sam back into the story and consider some of the big event points and investment costs that have occurred through time, Old Sam might tell us something like, well, the Louisiana Purchase land deal worked out ok, the Alaska thing was alright, you made a bundle with the cars, trucks, and roads, the fly boys and their spaceships cost me a fortune, but it paid off; and the way you clowns spend money, I know that the next one will make the flyboys' rockets and space program look like peanuts.

History proves that knowledge always leads to the next big event in time. In the recent past this term for knowledge has evolved into science and technology. Today, high speed electronic and digital switching/processing is still relatively new. It started to show up on the timeline around 1940. We can't say, just yet, where it will lead to. We know that it can produce electronic images of a universe that is beyond imagination. It has put some of the fly boys' machines on Mars, and as I recall it is making it possible for the first machine or probe to leave this planetary system and head for interstellar space, soon. This tells us the electronics, or electron and sub-atomic particle control, all appear to be leading man toward other parts of the universe. This short period of time, fifty years or so, or advanced space exploration, has just now produced a good starting point of things that will continue to be beyond imagination.

But there are some major economic problems here on earth that demand some attention, before we can all follow the electronic contraptions into the far reaches of the universe. Do not misunderstand the writer; I think it would be great if we could step into a device and dial up the right frequencies and field strengths to be transported to another planet or galaxy. One of the problems is tax funding pays for a major portion of the science, machinery and methods for such things to happen. Almost everyone who pays taxes, wants or needs to pay less tax in proportion to income, in order to make real advances in the quality of life and increase purchasing power or consumer spending. If we reduce taxes, that means funding cuts to advancement. History shows that in order to raise wages, raise taxes, increase value, and raise the standard of living, expansion and construction must steadily increase.

We could enter all of this into the timeline and common thread example, but it is getting crowded and would become confusing. We can continue using general history and indicators to form a few more ideas. Every so often wages double and taxes double. This is a result or product of the economic system. All of the ratios, percentages, and terminology that permit this to happen get very complicated. We can use a general statement that says, man's successful advance through time was due to continuous expansion, construction and spending that permitted and/or caused wages, taxes, and value to continually double through the years.

Time for a little science fiction. This is some odd logic that sounds like a possibility on one hand, and nuts on the other. But it is worth some thought, especially by the bored executives who need something to do. For most of us it is fairly easy to imagine that with all of our advances, especially those related to error-free computing systems, things in business are running like a clock. Changes happen once in a while that alter process one way or the other, then they quickly stabilize to a standard procedure. Things are lean and trim, and not much fat to cut; and most of the time it is even hard to get out on the golf course to make a decent deal anymore. Sound familiar?

Now suppose the boys in the back room were getting ready to hatch a plan that is different; it has the power to double and triple manufacturing and production orders in one day—but there is a catch or stipulation. It says: profit margins and prices will be frozen at current levels; and the new profit increases will be directed at doubling current wages from the bottom to the top; plus, double tax revenue. Then at some predetermined and agreed upon set of numbers or factors, the freeze on profit margins would be removed; possibly keeping in place price increase restrictions until a specified set of numbers are satisfied. After that, it's back to business as usual.

Suppose this idea came out of a G7 or G8 meeting (old Sam calls them the gang of seven or eight). The plan is to double wages and double tax payment to governments in a very short period of time, without raising or permitting the cost of living or inflation to increase until a certain set of numbers match. With other intentions of doubling the standard of living and reduction of debt. It is like a matter of finding the correct formula or equation; can you find it, can you make that work?

History indicates that there will be another "good" big event on the timeline of man's history. It will be larger than any in the past and smaller than the next big one going into the future; if the next one needs to have a larger price tag than space exploration, imagine how much that one will be worth. The world is a much smaller place these days; much of the old hate and conflict has faded. Suppose that before the next and biggest economic boom ever gets started, an agreement to double the standard of living of each nation was built in as a pre-requirement. Does man owe that to man, and the future? Is that economic fiction or economic science?

The combinations of man's science and science fiction have discussed building stations, bases, projects and mining camps on other planets and moons for a number of years, off and on. One point of view says that is only natural, because he can do it. Another point of view says that such ideas are driven or proposed due to economic need. The problem, at this point in time, is that such programs don't produce enough profit, value or wealth. Do not misunderstand me, man has gained data banks full of knowledge and made great advances with science and technology over the past fifty years or so due to the efforts of the Air Force and NASA. Here is the deal: suppose man had a choice to make about how he was going to provide for the economy of the future. Suppose he had to make a choice between building a base on Mars, or developing this method of construction in the coastal seas; especially, when we add in the idea that man will need to construct seawalls. How would you choose?

Here is another way to consider such a choice: let's bring old Sam back into the picture. Old Sam walks into the meeting and says, here is a trillion dollars for you boys to invest in the future. Do you want to build a base on Mars or do you want to develop the methods for construction in the sea? Well, if old Sam made an appearance with a trillion dollars, he wouldn't be asking anybody anything. He would be giving orders to develop the processes to save the country and help protect the rest of the world; you don't protect the country or the world from rising sea levels by building bases or projects on the moon or Mars. In other words, man would be investing Sam's trillion dollars into some new construction technology right here on Earth—in order to protect what already exists, creating profit and wealth in the process. If done right, then that process might actually create the tax base to pay for the projects on the moon, Mars and far beyond.

Before man commits himself to long-term, high-cost projects in space that do not produce large profits and employ millions of additional people, or provide protection from a severe threat, he must define the threat, problems, and the solutions related to rising sea levels right here on Earth. As far as financing the solutions, the fact that man has considered building projects on Mars tells us financial means are available in large quantity. So, it is simply a matter of directing such means to the place that can produce the most good or supply the largest need, that in turn can supply the largest profit. It's like Sam's boys back in the 1800's; they had to secure the mainland of the U.S. with investment into the western lands, before it made any sense to head to the North Pole and acquire those lands. Simple, but true.

Let's continue with this combination of science and economic fiction. Suppose that this idea of expansion and construction into the coastal seas would succeed and produce the excess amounts of wealth, profit, value (and seawalls) that it is supposed to; and suppose that an idea for doubling wages and taxes while holding prices was a success. Suppose that because of the extra hundreds of billions of new tax dollars pumped into space research, man was able to take those little sub-atomic particles that are faster than the speed of light and build his interstellar space scooters and transport machines. History seems to want to tell us these type of things fit in at some future point in time. Imagine the boys and girls finally finding another earth or two—maybe even unspoiled, with no human blood or bones buried in the ground—wouldn't that be something?

* * *

Finance, and One More Thing

Man has one more thing that should be considered, with regard to finance and this type of construction. This is in reference to materials, specifically rocks for making concrete. All across the United States, especially the western part of the country, man has material that was processed to recover and produce metals of all kinds. Some of this material is now classified as a problem, and is listed within the super-fund clean-up projects that have billions of dollars of cost attached to them.

Here is the deal. Most, if not all of these materials, if used to make concrete and sealed within the molecular structures that would form the large concrete foundation needed, would no longer be a hazard to man. In fact some of the so-called hazardous waste rock from the country's mining operations might be used to produce some of the strongest concrete ever made. The value of such material, especially the material that has high content of heavy metals, might even be greater than the value that the material was originally processed for. With a little study it might be discovered that the heavy metal waste rock should be blended or proportionately mixed into standard processes to produce a better concrete. If so, almost everyone would win or gain by using these materials to help produce the massive concrete structures needed for the demonstration project, or future projects.

We are talking about cubic miles of crushed rock that exists, and could be loaded onto trains and shipped to the project processing areas. Some of the hazardous clean-up projects are estimated to require additional taxpayer costs of billions, probably hundred of billions of dollars. The same material could be used safely, and actually create additional value and wealth that is real and meaningful. It must be considered that these tailings piles or muck piles (waste rock) exist all around the world.

* * *

The Chance of a Lifetime

We could say chances come in all shapes and sizes, and each person really needs to define their own. We can make an attempt to identify an opportunity as a chance of a lifetime, but some we just can't define or recognize.

I would like to think and believe the successful development of the types of construction these stories discuss would provide countless chances of a lifetime, for many people. When we consider the actual process of construction that involves sweat, dirt, busted knuckles, sore backs, etc., some of us might not consider that as a chance of a lifetime. Yet, some of us would. For those who don't, at least it could provide the means to pursue the type of life that might be desired. While up at the corporate level new construction orders are almost always welcomed by people that are, or might think that they are, living their chance of a lifetime.

The idea that man might actually be able to construct new cities that rise from the sea floor, and on a large scale, is one of those chances of a lifetime. Then, if we add the need for developing seawalls and that some of these methods might be in preparation for such need, well, it really is amazing.

The ocean is kind of like another world, in comparison to where most of us spend our lives. Different, unknown, and until the past fifty years or so, the depths could not be explored. About seventy percent of the land is covered by it; it only makes sense that man should use as much of it as possible. It is a lot of area that will provide a place for future man to expand and grow.

Anytime we mix the sea with the lower latitudes many of us, especially when we get closer to retirement, might want to think of tropical island paradises and tourism. If we combine such thought with manmade structures, one type of classification would be a manmade island. If we use the same structure, add office buildings and people to operate business and the support structure that forms a city, then a classification of a metropolitan area addition or suburb might apply. If such a structure were built off the coast of New York City, designed as an addition to Manhattan and Wall Street, with bridges, tunnels, and a Mag-Lev that connect to that area—it might be thought of as Wall Street East, or Wall Street Seaside, or maybe even Manhattan East. Such a design might relate more readily with the metro area as opposed to a tropical island paradise. So, it would depend on design and use.

Here in 2007 the chance of a lifetime, with any reality, might apply to a demonstration project. Maybe, comparable to the one we have discussed. It is a pretty well known fact that a large percentage of Old Sam's government jobs fall under the gravy train classification. Usually they pay really well; usually they are behind schedule. That indicates people are not pushed too hard; and usually when Old Sam is involved, things are engineered and built to the right specifications. When we start talking about a massive sized project that is largely funded by the government, experienced construction hands start to get itchy feet, experienced engineers and architects start drooling, financial controllers and politicians start licking their lips. To really appreciate the descriptive words in that last statement, you almost need a special understanding. Some of the most amazing moments of the writer's life have occurred when large construction companies start calling in the field crews from around the country and they start to assemble at a new project site; it's almost as though God has a special place and understanding for his construction hands, managers, and financiers.

Designers, architects, and engineers most readily appreciate jobs of this nature. It has to do with creativity, and large operating budgets that can permit time and expense to be less of a factor, that in turn permits designs and construction of higher quality and standards. Keeping this in mind makes it easier to understand how the chance of a lifetime might apply to the old pro with thirty years of experience or the kid just out of school. If we consider that the design of the demonstration project could include a World's Fair event, then some design and innovation could be incorporated and built based on that—it provides ideas for use, design and construction. A

World's Fair event supplies the reasons and excuses for using all of the newest, best, and most expensive products, designs, gizmos, whatchamacallits, thingumabobs, doodads, etc. Kind of like a dream come true for many of the engineers.

But most of the possibilities are far greater than one demonstration project. In an attempt to demonstrate some of the possible effects around the world, think of how the use of electrical power has changed the world in the past 100 years or so. If man can design the type of construction we are discussing, to have the same type of overall impact and effect, that continuously expands, evolves, and creates economic supply, we get a little more understanding of how important it might really be, and how many chances of a lifetime it might produce.

* * *

Considerations for Design

These next few pages discuss all types of things related to design and actual use, primarily for a demonstration project. It's easy to imagine that a large demonstration unit, as a model, might help set standards for a revolutionary method of construction for the world, with a chance of being something very special. If man could design the project as a showpiece that could draw millions of visitors annually, that would determine some of the needs and specifications. With that many visitors each year, transportation and accommodations are at the top of the list for considerations. If the project is a stand-alone city that caters to the travel and tourist industries, it would require all of the standard services and provisions of that type of a city, including government and operations personnel to attend to the needs of a city. This then requires the design of resident needs and services such as grocery stores, schools, churches, etc.

Some of the country's national parks have annual visitor numbers that are in the millions. Most of the parks have what are considered gateway towns and/or cities (each supports the other). Studying some of these can provide valuable insight into design, management, growth demand patterns, etc. It is possible the design of a demonstration project could involve ownership and management that compares with the national parks instead of business or private ownership—especially that of the primary structure. That could lead to long-term lease agreements for commercial business, services, etc.; or something to that effect. At some future point in time, if the project actually served as a beginning point of new construction design, it might be treated and preserved as one of the historical landmarks for future generations. But, on the other hand, the project might be treated like all of the other property that Uncle Sam has bought, sold, traded, fought for, and given away over the years—property that goes through various processes of development to form business, investment, free enterprise opportunities, etc. If such a project ever develops, we could be pretty sure that all of the possibilities will be studied to provide the best course of action.

We could expect such a project to develop about the same way regardless of who owns and or manages the structure. But one point of view really favors the idea of national park type management and control. One of the primary reasons is because it would readily open the door for the engineering control to the Army Corps of Engineers. If it can be built, they can get it built; if they control the design specifications, we can be almost positive that it will be right and provide the correct base standards for any future commercial construction. Another reason is because the Department of the Interior and the Park Service have the experience and are set up for managing most of old Sam's property.

* * *

A Demonstration Project

We have focused on some ideas about a demonstration project that could have a chance at serving as the major starting point for the type of construction being discussed. One of the things mentioned was size. Let's pull a number out of thin air—100 square miles—for something to work with. A one hundred square mile project built in the coastal sea might be thought of as a man-made island; it might be thought of as being a separate entity; or, it could be designed as a suburban addition to an existing city or metropolitan area. It might also be designed as a new, stand alone city (even high tech), or any combination of the three—it would all depend on the design. One hundred square miles is about 64,000 acres; one square mile is about 640 acres; one square acre is equal to about 43,560 square feet and that would measure about 208.7 feet x 208.7 feet. A 100 square-mile area does not sound very big to some of us. To appreciate the value that can be created with it we should think in terms of existing real estate that leases or sells at rates of thousands of dollars per square foot. Such considerations can provide a different perspective. One point of view and history want to tell us that if man designs the structure for high value, it will only be a matter of time until it pays for itself. To take it a step further, a brand new manmade island paradise that is almost hurricane-proof, located a few miles off shore in the warm sunny coastal waters of Florida that will pay for itself—that's not bad.

The key to this is proper design and engineering that includes engineering high value. Well, the world value has different meanings. In day to day living and the business world, we most readily think in terms of dollars and cents. But some values cannot be measured or calculated, especially when we get into the areas of morals, ethics, love, hate, good values for society, etc. Even some monetary values are almost impossible to measure accurately due to related interactions and their consequences. An example would be the combined costs and profits related to the automotive and transportation industries, if we were to attempt to include all of the operating costs, profit percentages, etc. A quick example: we could start with the wages that pay the taxes to build the highways; then all of the interaction, or the money changing hands, that starts with a paycheck and ends with a paycheck. We could add in automobile manufacturing and look at how far the purchase price spreads out—from the purchaser's wages, to finance, to insurance, to the factory, to the assembly line workers and his kid's tennis shoes. We could add oil and gas in the same way; rubber manufacturing; law enforcement; more taxes, junior's driver's education classes and the instructor's wages and his kid's tennis shoes; the highway workers and engineers. Then we could add in the grocery store and all of the costs that relate to transportation. Each area can be calculated separately but the in-between interactions and the effects are difficult to calculate; and they do have great value.

Another example are the institutions of higher learning; they have great book value. But if we try to calculate all the values produced by the actions of those instructed, it is next to impossible. Similar, yet different, is the value of the national park system; we attempt to combine all of the measurable values with those that cannot be measured. Love has value that cannot be measured; its effects on people can't be measured. The old saying tells us that it makes the world go around; that's worth something. Yet, we can't take it into the grocery store and purchase a loaf of bread with it. These examples tell us of values that can't be measured, yet man's actions, designs, and engineering produce such values. One of the keys to the production and promotion of measurable and unmeasurable values is proper engineering and design—that's the bottom line of these last couple of paragraphs.

If or when man successfully builds his first new manmade island, suburb, or city that rises up out of the sea, even before the city is built on top of it, it would provide the positive proof of the possibility of construction and growth that would be beyond imagination. The process of building the city on the first foundation would be like the icing on the cake—the city would be for man's enjoyment and protection. Much of the enjoyment and protection would be derived from monetary value; from the kid selling hot dogs at football games to the financial stability of a nation. The foundation structure will be the hard part.

Before man can sell the idea to business and government, he would need to have some of his designs prepared in advance. How would you build it? We can add almost anything we want to into the design. With that in mind, this manmade paradise, this twenty-first century wonder of the world, should have it all and more! One hundred square miles can provide the area for a considerable amount of construction. If you are still working or getting ready to start a career, think in terms of the perfect job; if you are retired or near retirement, think retirement bliss; if you are a corporate manager thinking of new headquarters and building, this is the place; thinking of the perfect vacation spot?—this could have it all.

First, let's discuss the obvious, the sea! We could fill a couple of books with just ideas related to the sea. Let's look at a couple of pages. Some people do not require much to be content and enjoy life—this seems especially true when one is located next to the sea. Folks like this might only require a nice view of the ocean from the balcony of their condominium, hotel, apartment, or the backyard of a town home; along with a safe and secure place to do the shopping. Other folks like to get in the water with things that range from inner-tubes to jet skis, sailboats, fishing boats, luxury yachts, to cruise line ships, to aircraft carriers. A while back I even heard of a company that builds personal use submarines, so accommodations for eccentric retirees who like to jump in the sub and go for an afternoon cruise would need to be available. Of course all of these would require sales, service, and support personnel, facilities, etc. These types of things provide opportunities for many people.

Old Uncle Sam would have a considerable amount invested in such a project. It is easy to imagine that he would require coast guard and naval facilities for a few of his salty old sea dogs, NOAA, and a few others. At this project military and government duty would probably fall under the gravy-train classification. Thinking of duty that deals with the public and kids for educational, monitoring, and emergency basis duty.

If we think of corporate business and some of the powerful executives who might want to move a few company business complexes to a new high tech city, design that permits flexibility for construction and preferences need to be considered. A project such as this might be thought of as being unique or one of a kind; and far above average with design geared toward first class service, accommodations, travel, etc. Costs to build such a project would be high. Wealthy folks who are in the upper classes do not need low-cost or average. Many of these people enjoy paying high and seemingly excessive costs, in return for high quality products and services; this is good, because it helps provide opportunity, pay wages—in addition to creating demand for a few of the finer things in life.

I would expect a large portion of such a project to be designed with some of these thoughts in mind; and also provide the proper percentages of accommodations for the average family, who does most of the work, and might wish to travel there on vacation. Ideally, overall design should be directed toward raising the standard of living and quality of life.

* * *

Hurricane-Proof Construction

Can man build a hurricane-proof city that might have a population of a few hundred thousand people? The answer for man in 2007 is yes but . . . The biggest but in this answer relates to cost. The next but relates to a percentage of construction, form, function, protection and the unexpected that cannot be designed, built or utilized in a hurricane-proof form.

The primary supports that rise from the sea floor, the deck structure and the core structure of the primary buildings would, or should, be designed with enough strength and mass that there would never be a need for concern of the forces of the wind or water causing any damage, period.

But the secondary construction that builds the actual city is a different matter. Fifty-one percent of the secondary construction, buildings and protective designs, might provide hurricane wind protection for fifty-one percent of everything that occupies space on the structure. That number would indicate forty-nine percent could be susceptible to damage or being blown into the sea. Engineering design could make these numbers show almost any percentage, except 100 percent storm-proof. There will always be a percentage that cannot be protected from storm damage. If we were to attempt to place percentage and dollar amounts on the entire structure that could not be protected, dollar amounts might produce numbers that show seventy-five percent of the value, hurricane-proof, and twenty-five percent susceptible to damage. Understand that these numbers are best guesses. It all depends on engineering design and costs.

The only way to provide storm-proof construction is by using concrete, metal, protective seals and covers to surround or contain everything, defy the wind and keep the water out. The almost indestructible construction might be thought of being built like a battleship. But, exterior coverings, sidings, ornamental stuff, and all of the frilly things that are attractive and pleasing to the eye, could supply everything necessary to provide a more than acceptable appearance, and function. But most of the frilly doodads, siding, trim, etc., would probably end up belonging to the sea with each large storm. Insurance terminology sometimes refers to such things as acceptable loss that is easily, and often desirably, replaced.

Man has the know-how, tools, and money that will allow him to continue to provide safety and protection against these powerful forces of nature. History tells us he must, and will, continue to develop these. When we get into areas of protective design that include things like automatic motorized systems that close window shutters and doors when wind gusts hit 50 mph or so, we can realize how recent advances can produce storm-proof construction that prevents destruction. When man can prevent loss, this indirectly tells us the savings can be used for investment into new high quality construction; as opposed to the continuous and neverending process of cleaning up and rebuilding every time the wind blows. Yes, the initial costs are higher, but the real value and profit creation is derived from preventing destruction, not the continuous process of building something, then permitting it to be torn down or destroyed, then again rebuilding. The repeating build, destroy, rebuild cycle does help supply and produce the economic demands; but man has a chance at creating fundamental changes that will permit him to continuously construct new long lasting storm-proof designs of great value, derived from these types of construction and expansion methods.

Consider this: the production processes that would be demanded for the high quantities of materials needed to construct the foundation structures would be huge; they would be something from out of this world; they are science fiction material. In the same way space flight, "was" science fiction. To give us an idea of such processes we could think of all of the concrete that exists today, in some of our largest cities. Try to imagine the processes that might be required to produce an equal amount of that concrete, in a matter of a few years and working on processes that would further reduce it to a matter of months—at some future point in time. That would be very serious high capacity production, even in the earliest stages of development. Think of processes that might be able to pump concrete into the forms for large building structures in a matter of hours, if not minutes! Because of such capacities, costs could be reduced to realistic and manageable levels providing for a new preferred method of construction that would permit the hurricane-proof construction we are discussing. It really is an amazing concept, one of those tall tales almost beyond imagination. But look around you at all of the tall tales of the not-so-distant past that really did come true!!!

Here's another thing about construction materials, and how the use of concrete and metal will continue to increase. In a way, it is easy for us to imagine a few things that relate to how the future will develop in the next 300 years or so—if we compare today with the 1700s. For example, forest and wood products. Try to imagine the consumption and value increase rates of these products over the past 300 years. In another 300 years, around the year 2300, costs will increase and supplies will be much less plentiful than they are today. That tells us more and more replacement materials will be invented and developed to supply man's needs. Such as plastics, metal alloys, steel, composites, concrete, etc. If we look way out there, it's easy to imagine metal framing will replace wood as the normal building material for homes, and that the use of concrete will increasingly grow. In the 1700s, an acre of ground full of virgin timber could be purchased for a couple of dollars; in 2007, a good straight piece of wood, nominal size of two inches thick, four inches wide, and eight feet long, can cost about two dollars. Using comparative rate and cost increases, in 2300 a good straight

2x4 might cost a few hundred dollars or so . . . it's hard to say.

With the right design, the demonstration project could continue to be, almost, fully functional in a full force hurricane; even while all of the things not protected or bolted down might blow into the sea. The utility systems (even the large generator plants), some of the walkways, roadways, parking areas, dry dock storage areas, subway systems could be placed in sub-levels to the structure where they would be surrounded and protected by the primary concrete and steel structure. One of the exceptions would be the airport; man would need to keep the planes in the hangars until it blew over. It would not be cheap, but it could be secure! The kind of construction that would permit the engineers and man to give old windy the finger when those 200 mph winds blow.

* * *

This portion of the story presents a few ideas about how man might design and build the flat deck structure that would be built on top of the primary supports that rise up from the sea floor. The fictional demonstration project we have discussed is 100 square miles—that is ten miles long and ten miles wide. If you could ride a bicycle all the way around the outside perimeter at an average speed of ten miles per hour it would take about 3.9 hours to go around one time. If you could drive a car around the perimeter at sixty miles per hour it would take about 39.9 minutes, so about forty miles. To provide one more reference for size and material think of the interstate highways that cross the country. If we used ten mile sections and placed them side by side to form a square the same size, ten miles x ten miles, it would take about 13,200 miles of four lane interstate. The structure would be large, but not too big.

Once the deck structure would be completed, all of the additional construction would take place on top of it, below it, or inside of it. The office buildings, the condominiums, the airport, the football stadium and golf courses could all be built on top of it. It would be real important, and there are a few different ways they might be designed and built. It would depend upon the calculations the engineers come up with. The deck would have to support any load or force applied to it. When we start thinking about landing loaded passenger jets on it and having stadiums with 100,000 fans stomping their feet after a home team touchdown—we get an idea of just how tough and strong it must be. In some places the deck might be solid reinforced concrete that is ten feet thick, twenty-five feet thick, thirty-five feet thick, maybe even fifty feet thick. In some places it might have sub-levels below the deck surface that could provide for all types of things, such as a subway system, utility systems, dry docks, storm protection for people, parking areas, even roadways. One point of view says we could think of the bottom side of the deck as a multiple level basement that could be almost ten miles wide and ten miles long.

The sub-levels would offer other very valuable possibilities. Consider this: suppose this entire structure was a self-contained city, located a few miles offshore from one of the large East Coast cities. It might have a population of a few hundred thousand people, with most of the major construction completed and everything working well. Now let's think in terms of civil defense shelters and protection. Suppose that there are all kinds of transportation routes between this offshore city and the mainland city. There might be two or three sets of subway tracks that run in tunnels, one or two auto and truck tunnels, one or two bridges, maybe even a high speed mag-lev or two suspended overhead on towers. With the proper design, the sub-levels of the structure could provide refuge areas for millions of people, if ever needed. The area could be designed with filtered air systems; if needed, lead liners could provide additional protection. The military would probably have a station and dock area; commercial shipping would probably have dock areas; both of these could serve as evacuation points, if needed. There would probably be a couple of airports on the structure; they would serve as other evacuation points. In addition, the reverse could apply. It could also be used as a reception area for masses that might have had to evacuate from a different location, those who needed a place to go. These are ideas that must be considered if the idea ever gets close to a drawing board. Old Sam might even be a little more interested in the development, if such civil defense protection could be built into the designs.

Consider the height above the water. First, engineering would need to determine the proper distance, we can suppose. We wouldn't want storm waves to reach the deck structure. That says it needs to be up there, quite-a-ways. Some places might only require one hundred feet or so. Other locations might require 300 feet or so. Man has all types of data to provide average numbers, and the extreme numbers required to determine the correct height. Proper designs would provide resistance to tidal or tsunami waves, eliminating or reducing damage caused by such events (in addition to early warning systems). If we take into consideration the discussion about the need for seawall systems, it will affect the height requirements. If these manmade islands and cities were built inside the seawall protection systems, rising sea-levels would not be much of a factor. But, if they were built outside such systems, they would need to be high enough to handle the rising sea-levels.

* * *

Miscellaneous "Stuff"

We discussed that a large demonstration project, that would cater to the travel and tourism industries, should have it all! Let's briefly discuss a few ideas that might apply to this manmade paradise. One of the first things that comes to mind involves the transportation systems that would be needed and desired. Such a demonstration project, located a few miles offshore, that attracts millions of visitors annually, would need about any type of transportation that man can dream up. Because international travel could be a large percentage of the activity, all of the things that relate to that must be considered. Such as an international airport with customs and immigration provisions; the same would apply to cruise-ship port and docking facilities. Then, if we consider connections to the mainland, it is easy to imagine some of the types of bridge and tunnel systems that might develop for autos, trucks, subway trains, even the new mag-lev trains. Add in the big high speed ferry boats. This provides a few indications of the types of opportunities that could be created with the development of such a project.

One point of view, simple and basic, says, the rest of it is the stuff that dreams are made of. Such as providing accommodations and activities for all of the millions of visitors and building the homes, services, and businesses for the residents.

Many millions of people enjoy sporting and gaming events. People from all walks of life enjoy the process of traveling to such events. Good and positive effects, produced by the large high capacity professorial events, should all be considered in the design of a demonstration project. Professional baseball, football, basketball, golf, and (a favorite of the writer) Indy car racing—just to name a few.

Most of us could agree such events would be pretty nice in a warm coastal sea location during good weather. But when the weather turns, it wouldn't be so nice. To insure against foul weather for the protection of 100,000 fans in a stadium, with the proper design, an aerodynamic shape would permit the wind to slip over and around the structure. If additional measures were taken, the structure could be made to withstand old windy's 200 mile per hour winds. In addition to this, a few other measures could make such stadiums a very valuable asset for civil defense measures of all types. As far as protection for the fans of PGA golf events and auto racing (and the cars) the use of sub-level areas or floors could take care of that. One of the great things about building the whole thing brand new and from scratch, is that man could design almost anything into it; to make it better, safer, and more desirable.

Now the writer mentioned something about raising dairy cattle in high-rise structures; maybe we should clarify this. To come close to any kind of appreciation for such an idea, it is easier if we apply some of that science fiction thought that takes us hundreds of years into the future. If the construction method would become a success and develop into large areas, that might compare to the area of a small state—we can be pretty sure that the spirit of some old cowboy, cattle baron, or milk producer is going to want to raise some cows.

As far as the demonstration project goes, the idea is geared toward education and experimentation, related to futuristic need that would directly relate to these types of structures; and might be developed in cooperation between groups such as future farmers of America, the 4 H, the boy and girl scouts, and any agricultural school that might want to be involved. Plus the USDA, due to the knowledge base and deep pockets of funding that could help to build something special and valuable. We can stretch this type of thought a little further, leading to areas that include the development of soil from sea sediment, cattle waste, and ash from the new low pollution coal-fired power plants that should be included. Man should also consider hydroponic type production of vegetables. Because space would be so valuable, things would need to go into high-rise structures. Then there is fish farming, and all types of food production that would be unique to this type of construction. Besides that, it would be good for the kids.

Conclusion

One to Grow On is a story about making money: it is a story about money being made for people; it is a story about money being made for people to make money with, so that they might live in freedom without the need to create war in order to supply the economic needs of man as he works on building the future. In addition to this, it is about protection from the threat of increasing sea levels. For the writer, it was almost as though someone hit him with a brick and said, hey dummy, develop these methods so that man can build the sea walls is what this is all about.

These thoughts of building cities that rise up out of the sea, manmade islands and peninsulas, and 100,000 miles of seawall and lock structures, for many of us, are beyond consideration. Could be that it is just too late in life or that we are simply not interested. Yet just the opposite will be true. This refers to the people who will help make such things happen, and become real. For such people, it might not require much study or research to get a feel for how real some of these ideas and comparisons are.

Well, the writer had about a half of a dozen longwinded pages to close this story, but the latest revision says that we covered the important "stuff," and that there is no need for them. So . . . let's use one of the old sayings to wrap this up. It is the one that goes something like this: It is true, that there are moments in time when fact really is stranger than fiction.

Water for the West
(Story Number Two)

Water for the West is a short story about water. When it comes to building the future, fresh water supplies will continue to be one of the most important factors; shortages exist now. This story discusses the problems of and solutions for our primary water supply, especially in the Western U.S. The solution is the objective and focus point.

The solution might be found in a new process, designed to convert sea water into de-mineralized fresh water. The process could use the natural forces of nature, or atmospheric forces, to pull or push sea water through a filtering system; by taking advantage of the difference in elevation between the surface of the sea and inland locations that have elevations below sea-level. The power to make this process function would be produced by water flowing or falling to a lower elevation. Two types of natural atmospheric forces are produced when water moves. Pressure and vacuum (or suction); said another way, push or pull. These are forces that make filter process function.

Basic engineering principles tell us man can design a method of creating these two forces that can be powerful enough to pull or push vast quantities of sea water through a filter system that does not require electro-mechanical power to make the primary filter action function. The costs related to such a process could be reduced to levels that make some kind of sense, and man might be permitted to produce all of the water he needs.

Basic science tells us, if man were to install a large pipe or tunnel between the Pacific Ocean and Death Valley, the natural atmospheric forces would force water to move through the pipe, downhill to the lower elevation at Death Valley. We could think of it as free flowing water that does not require pumping costs. Basic science also tells us the inlet end of such a pipe or tunnel would have some very powerful suction force. The same basic science tells us if man designs and installs the proper type of filter at that point, he will have clean water flowing on the lower end. So that would produce desalted or filtered free flowing water running downhill from the ocean. This simple example demonstrates how forces thought of as vacuum, suction, or pull could be used to make a filter work.

Inside of that same pipe, because of the flowing water, another force would be created—the force of pressure (or push). Simple science tells us if man were to design the proper filter and install it in the pipe, water would be forced through the filter and could provide clean water on the discharge end of the pipe. This gives us the basic principle of how a filtering process could be created using the force called pressure.

This provides the bottom line, basic theory of a solution for one of man's neverending battles related to water supplies. One of the major problems, in an attempt to utilize this type of design, is that there are very limited amounts of inland areas that have elevations far enough below sea-level to permit the installation of such systems. The writer's basic research has found only three locations in the U.S. that have elevations that might permit conventional use of this design; there could be more. If man is to make use of this type of design, it will require investment, and expert engineering to make the correct stuff.

With just a small amount of interest in water, combined with the understanding of these basic principles, we can sit back and ponder the possibilities and opportunities that might be created with designs such as these. This is one of those stories, in a short version, that combines a little science. When it comes to actual use or development, it is like all of the science fiction of the past that produced today. We could apply a couple of the old sayings that tell us things like, only time will tell, and the other one that says it will only be a matter of time. Sometimes wording like that helps permit the imagination to wander around and dream up a few things.

Water is important stuff. The human body is said to be about fifty to seventy-five percent water. Water is said to be a liquid that is a combination of oxygen and hydrogen, and weighs about 8.3 pounds per gallon. That says a 150 pound body is around nine to thirteen gallons of water. Here is a little secret about shedding weight; you can sweat it out. In past years, the writer has worked in jobs that were physically demanding, and the men who did the most work and perspired the most usually never had a weight problem, ate like horses and drank beer by the gallon. This is sort of trivial in relation to the story, but some might find it interesting and it is true. Man can survive days without food, but he won't last long without water. Fundamentally, that almost makes it the most important thing there is. Then, if thoughts wander and we start thinking about how many toilets need to flush every day, and how many acre feet are required to fill up the lakes to float the boats and keep the fish happy, well, it is pretty important stuff.

The stories, history, indicators, and price of water tell us man has some big problems waiting for him in the future, with regards to water. Especially in the western half of the country. Shrinking water supplies cause problems of many types. Such as shortages in irrigation water supplies that produce food crops; supplies for existing towns, cities, ranchers and farmers, industries; supplies necessary for growth and expansion; and even the wildlife, fish, and craw-dads.

A few fee of snow. It's like this: a few feet of snow on top of the Rocky Mountain Range holds a primary key to man's survival; and provides one of the most important things that he cannot live without, water. We don't know the extent of the effects of the changing weather patterns, or the effects of the changes increased solar activity will produce. But one thing is perfectly clear: if it should fail to snow enough, up on top of the Rockies, to produce the required amounts of water needed to supply all of the demands; or, if man does not develop a way to inexpensively filter large quantities of sea water and get it flowing into the rivers, there will be real trouble. It won't be just a localized or a regional problem—that few feet of snow, more importantly the lack of it, has the power to bring Wall Street and governments to their knees. A water shortage crisis in the Western U.S. could cause economic disaster around the world; a solution must be prepared.

The large lake and dam projects that exist are reported as not being able to supply the amount required, and some storage supplies are far below 100 percent. The mighty Colorado River almost dribbles through the deserts of the Southwest. Thousands of miles of tunnels, pipes, and aqueducts move water from place to place; and the shortages continue to increase. So how will man deal with this?

All types of studies and proposals are constantly in the process, yet there are still shortages. The writer decided to do a study of his own that, to his knowledge, has not happened, or been openly discussed. To come up with the idea for this solution the writer had to study history and almost go back in time to when the first mining prospectors, who gave Death Valley its name, were living and dying in that area. That is a long way back. Some of the hardships and terrible conditions they had to endure are hard to even imagine. But some of them made it and were able to open up that part of the country. Hotter than the hubs of hell, like the old-timers used to say, is a good way to describe the place during the heat of the summer. We can figure quite a few of those old miners worked themselves to death trying to dig a little gold and silver out of those desert mountains.

Death Valley is reported to be about 280 feet below sea-level; man and the West need water. If we stand in Death Valley and look to the west, it is about 300 miles to the ocean. It has all of the water man could ever use if he could get the salt, various minerals, and contaminates out of it. To understand how this process would work, you need to think of the difference in elevation between Death Valley and the surface of the ocean; that is about a 275-foot drop. With all of man's designs over the years, he should have no problem designing a very cost-effective (cheap!) filter system that can operate on either the vacuum created by the syphon action or by using pressure created by the weight of the falling water. In addition, a few turbines for hydro-electric power could be installed in that 200 and some odd foot drop in elevation, and probably supply all of the electrical power needed to make this type of design function, with plenty to spare for use at other locations; most importantly, military installations.

One of the next tricks to all of this, is getting the water over to Lake Mead. That would allow for the diversion of upstream water to help replenish low storage supplies in all the water-holes between there and the headwaters of the Colorado River. Then any excess capacity could be moved into systems such as the Big Thompson project to help supply water demand on the eastern slope of the Rocky Mountain range. The indications are, there is a chance to add some amount of water supply to the Plat River that eventually dumps into the Mississippi; then, depending on the miracles the engineers can produce, the project might add capacity to the Green River that comes out of Wyoming; even possible additions to the Arkansas and Rio Grande, however small they might be. It could be thought of as opening the tap or turning on the spigot that connects to the ocean. Then, from Lake Mead downstream, the systems that supply California and Arizona should also have plenty of supply; if not, keep adding tunnels, pipes and filter systems till they do. Well, that summoned it all up pretty quick. One thing to remember, it is all easier said than done. But, if it can be done and man decides to make it work, there is almost no doubt that it is going to happen.

Now, we could go on and on with thoughts that relate to developing such a method of creating new water supplies, but we won't do that with this story. We can discuss a few more things. Almost everything that would relate to such a project would be good for everyone; almost no one loses with a project such as this—now or in the future. Great amounts of economic supply could be created, starting with engineering, through construction, to the excess water supply capacities that could help to lower the costs of anything that has to do with the high costs and short supplies of water. Possibilities exist for the creation of water flows that could be used to create new lakes and reservoirs. Those could provide additional expansion and construction opportunities. The list goes on and on.

If this idea would work so well, why hasn't someone already developed it? The best answer to this question is that, somehow, this is one of the those ideas that was overlooked or possibly thought of, but considered too expensive—it is hard to say for sure. One very, very important thing to keep in mind—the cost to construct such a project would be peanuts, compared to the economic collapse that would occur with a water shortage crisis.

One other thing to consider is this: in the discussion about the Hoover Dam, the words, "a bargain at any price" were used. Man might want to try and apply that type of reasoning to this project. Along these same lines of thought are the ideas of important investments in the future; these also apply.

Construction. A few quick notes. Even if man can make such a process work, the way we have discussed, there is another problem; Death Valley is a National Park. That tells us an act of Congress would be required to do any type of major construction in the area. If man can benefit from the development, as we discussed, we can almost guarantee the project would be permitted. Death Valley is about 135 miles long and five to fifteen miles wide, according to some of the numbers. That would make a very large fresh water lake. Many people would consider the ability to create a water hole that big in the middle of the desert, maybe as a gift or a blessing. If we could ask the old miners who opened up that area, their thoughts on the idea, they might tell us something like this: son, any day above ground is a good day—if you can flood that stinking valley and create a lake covered with boats and pretty girls in swim suits, and build homes along the shores, build a dam and flood it.

Driving tunnels between the valley that would connect directly to the sea could be risky, especially in earthquake prone areas; it could get out of control real fast if one of the big shakes opened the ground. It would require systems that could be sealed quickly, in the event of a large geological event. The design requires safeguards and good professional engineering. By installing the filter systems in the sea or at the seashore and using a vacuum action to draw the water through the filter system, all of the materials and minerals removed from the water could be returned to the sea, as opposed to filters on the discharge end that would require costly handling and processing.

One more thing: on the other side of the world, or the earth, is another big sinkhole that is below sea level, The Dead Sea. It is odd that both of these holes or depressions in the earth have names that refer to death. If we permit our thoughts to wander toward God and religious teachings, combined with economic need, we can develop a few more ideas to think about. The entire region around the Dead Sea has great religious value. History tells us civilization started in that part of the world. Beyond all of that, the religious teachings tell us that the area has or had a special relationship to God. With references to being God's footstool, and having a city whose name has a definition of the foundation of peace, Jerusalem. Common sense tells us, don't make trouble at the foundation of God's peace; and if he wants to receive the blessings of God he should do all that he can to help maintain the good in place on earth that has a definition of the foundation of peace.

If we come back down to earth and get real, truthful,

and honest about things, economics holds a large portion of the power that helps to provide for peace and right ways of living. This indirectly tells us that providing good economic means and supply is one of the primary keys in making this world function in ways that provide for right living. Common sense and understanding tell us when people have work and a way to produce income, it helps produce good things as opposed to bad things; and when people work and are in pursuit of happiness, they do not have much time or desire to cause trouble or disturb the foundation of God's peace.

The bottom line to all of this: Man, at some point in time, should consider trying to develop the method of converting sea water into fresh water by taking advantage of the difference in elevation between the Dead Sea and the ocean. Endless numbers of projects for producing economic supply could be engineered and developed with this. This indirectly tells us man would then be helping to maintain God's foundation of peace—located only a few miles away. If we apply some modern day logic to this, one point of view will tell us, being on the proper terms with God is a good thing, as opposed to getting on his bad side.

Death Valley and the Dead Sea. Man, with knowledge that is said to be a gift from God, has developed the ability to turn these two holes in the deserts into water-holes full of life and opportunity. Will he?

Maybe it is just another chance of a lifetime.

The Other Future
(Story Number Three)

Man is heading into the future with constantly increasing rates of speed, that relate to almost everything. The consequences of his actions will happen and be altered at those same, ever increasing rates; both good and bad. If we apply the spirit and God into some of the equations that measure rates of change, some of the results can be gauged or measured mathematically. Yet when we move into the areas of scientific and mathematical absolutes, those things of a spiritual nature most often cannot be identified, let alone measured. But the soul of man is spiritual and the effects are a part of everything that he does in life, in some way. This is the direction for thought in this story.

Man, God, and the power of influence between the two is part of the spirit; all combined it can be very powerful stuff. There is an old saying that tells us, there are no atheists in a fox hole. This is representative of the power and effect of the spirit. In scientific and mechanical procedure, and political and diplomatic affairs, the rules of business and protocol do not specify the spirit of the soul as needing or having any effect, influence or recognition. Yet at the critical moments or very decisive moments in life, many of us, from all walks of life, reach into the spiritual soul of man and ask God for help and guidance; and that is ok, because it is a personal matter between man and God. History has proven and demonstrated this all through time—and this, if nothing else, will always be the difference between man and the machines and processes that attempt to imitate human life.

At those moments in time that compare with being in a fox hole, when man is between life and death, many of us ask God to save our lives or take our soul in death. Sometimes during moments such as these, it is like we have a direct connection with God and nothing else—no interference. The power of the spirit, at moments such as these, can be of great help and benefit to men, women, and children. One of the interesting parts of life is that some people can cultivate that type of spiritual influence between man and God for help in day to day living.

Almost any time we ponder the future, thoughts of space exploration are going to make themselves known. They go hand in hand and it is hard to think of one without thinking of the other. Now, if we think of the time when man could head off into interplanetary space, we think of all types of things. But there will be times when man gets himself between life and death, and right and wrong—when there is not a deciding factor that is determined by a science fact or a mathematical absolute. We can be pretty sure that both good and evil exist in various forms throughout the known and unknown universe. It is kind of like this: if I send my kids off into interstellar space travel they should know some things about the unknown supernatural power of God and the spirit; to draw from at those moments when needed. It helps provide God's influence, power, and protection as part of their guidance package—kind of like having God ride shotgun. Sometimes it can provide the edge needed. The same way that prayer can help when kids are sent off to war.

We do not know what awaits man in space; maybe he will meet up with an extraterrestrial. Could be, if he opens the wrong door it would destroy the world as we know it. But, if the influence of God's guidance is involved, we can be pretty sure that it will be of assistance—it might even help to find the right place to go.

Everything man has done in the past has brought us to this point in time. Just as this point in time is a stepping stone or preparation for going into the future. The past fifty years of space technology research can be described in one word: *miraculous.* The economic benefits can be described with that same word. The speed or rates of change, same word. But man is just getting started, just getting warmed up! Fifty years is like the blink of an eye or a drop in the bucket. History wants to tell us many more great things are going to happen and they are going to happen quick!

The near future should hold some interesting stuff, with regard to space flight. For most of us, the byproducts of the technology that are produced affects us more than the actual flight programs. Talks and plans for outposts on the moon and Mars have been going on for years; some day they are almost sure to happen. If or when such things happen, well, that much more science and opportunity can be expected to develop.

One thing that is especially significant, related to space, is the ability to detect, deflect, and destroy meteors and asteroids. I think the kids would have a ball setting up stations and equipment on the planets and moons to take care of such matters, as long as it is managed properly. The value of deflecting a large meteor might be priceless. History indicates that man will need to use all of the technology and abilities that he develops; so it sounds like a base on Mars at some point in time.

It is a funny thing. Man started out as hunter and gatherer, wandering around looking for stuff and at stuff—he still does the same thing, even in space travel.

We have to wonder, man, over all, has done a pretty fair job of building the Earth. Now, is it the job of man from Earth alone, to explore and develop the remainder of the universe—or are there really other civilizations out there? When we have time, it is interesting to think of the possibilities because there is a chance for the existence of other planets and civilizations that compare with Earth. Imagine finding one that was more advanced than Earth, one that might have the answers for all of the questions about today's and tomorrow's science fiction. Wouldn't that be something?

In the meantime, man must continue to hunt, gather and develop Earth. He will be in need of large sums of capital to fund everything, especially exploration ventures into the vast areas of space. In the story, One To Grow On, we have a few ideas that represent types of development and expansion that could be required. They are indications of man's needs that will move him forward through those future points in time.

Science teaches us with facts and numbers that provide theories. The Bible teaches man with words that were inspired by the supernatural or unknown power of God. There is a conflict between the two, and the kids always get the short end of the stick. Hopefully, someday, science and religion will be able to give the kids a straight answer. It might tell them the unknown supernatural power of God did in fact create the universe with his big bang; and that if evolution was a part of his process, in the creation of humans, it was only part of God's creative process. Any evolutionary process ended with Adam and Eve—when God gave them a soul, spirit, and counted the hairs on their heads.

There are special warnings about the soul, the spirit, and the holy spirit—these parts of man that are way beyond the reach of science. One tells us to protect the soul. Another tells us to fear the one who can destroy the soul. The soul and the spirit are special qualities God gave man. Always protect them.

Epilogue
Building the Future

Sometimes, when we begin to think of islands and permit our thoughts to wander, we can reference the stories of the Atlantis and maybe even drift into thoughts of Elysium's Islands of the Blessed—that's ok, they are nice things to think about. The writer gave some thought toward these while working on this book, and a few interesting thoughts came to mind. The old stories tell us Atlantis was a mythical city that disappeared, and was supposed to rise up from the sea, sometime in the future. If so, we might consider that it probably started reappearing with the construction of Venice; and continued to grow and expand with the coastal cities of the world that have been created since then. The idea of the Islands of the Blessed is also interesting. If we look around the world at all of the island nations, and the small islands with few inhabitants, we would probably find if we could interview a few people who call them home, they would tell us they are blessed. In a way, the mythical tales have already come true—starting many years ago, and continuing to expand to this day. Well . . . the writer found the thought interesting.

When it comes to building the future, history demonstrates that man in 2007 is just laying the groundwork or the foundation for portions of the future that will develop with time. The far distant future—we can only imagine and God only knows. But, for the near future, man has some great advantages provided by his increased knowledge, statistical data, past experience, computerized electronic control, etc. Things are not like the old days, and he has everything necessary to provide, grow, and expand in all areas. From one point of view, especially in the United States, we can say a large part of the hard work, development, and engineering is completed—from here on out, all man needs to do is build, grow, refine, and educate.

Man now has the ability to take the facts, statistics, and probabilities, process them, and form accurate estimates and predictions for all types of things.

The processing power provided by the computer systems give man tools and advantages that were only imagined and dreamed of just a few years ago! It wasn't very long ago when the world's engineers made their calculations with pencils, paper, and slide rulers! The ability to design, engineer, manage, manufacture, finance, invest, educate, and govern has never been easier or more open to so many people. Creating demand for all of this ability will be one of the keys, if not "the" key, to proceeding into the future—with prosperity and peace. Especially as the populations continue to double and automation increases.

The ideas and thoughts presented in this work are explained and, for the most part, easy to understand—even though building cities above the sea sounds a little wacky and far-fetched. There are experts around the world who study and research just about everything we can think of, and more, as part of the process for providing the needs of today, tomorrow, and beyond. New ideas and inventions are always on the drawing board and in the process of development.

The ideas and discussion of rising sea-levels demand serious consideration; if not properly prepared for in advance, the devastation would be terrible, worldwide. In the recent past there have been discussions of attempting to control the rising temperature of the earth by combinations of methods. The causes of rising temperatures, in relation to the rates of polar ice melting and sea level rise, are most likely far beyond man's ability to control. This means man will be demanded to begin seawall and lock construction far in advance of the time they will be needed. The option would be to abandon the coastal cities and move inland. Most of the world's largest cities are coastal ports; it would not make any sense for man to allow the cities that have taken hundreds and thousands of years to build, to become permanently flooded.

Insurance and business use the term, act of God, to classify certain events. We can think of the rising sea-levels as an act of God; one that man has the ability to manage in a way that could provide the world's largest economic boom, economic stimulus, economic supply source, or work program—ever!

Properly engineered and managed, this holds the power to create and secure the financial and economic generators that will be needed . . . for a very, very long time. As an added bonus, man should be able to eliminate the need for future wars.

Words are a funny thing. Some are very simple and easy to understand, and some are very complex and difficult to understand. People speak, write, and arrange words for one primary reason, to communicate. The motivation behind the need to communicate covers the range from a to z. Words can be very powerful at times, and sometimes they produce little or nothing at all. As an example, these words are transmitted using paper, ink, and the electrical impulses, using things called photons and neurons that move between the ink on the paper, your eyes, and your mind. For some of us, that thought can open a door within the thought process creating a sense of amazement. For others, they are simply words on paper that have meaning, but they don't produce much thought or effect. There is no clear right or wrong, it is just a difference in the thought process.

Words can make fortunes, and they can lose fortunes. Words help in the process of creating and protecting life, and they have the power to kill, almost instantly; that is how powerful and valuable some words can be. It is a strange perspective; in our day-to-day living we do not usually have the need or time to judge the power of words in this manner. They can produce good and, on the other hand, they can produce an equal or greater amount of evil. Only God knows how many words each of us will hear, read, speak and write in a lifetime. Some people speak plain, clear, and mean almost exactly what they say—even when using those big fifty-cent words. Some of the others speak plain, clear, and seem to make sense or sound good—but their words contain many lies or untruths. Some of them would rate as expert professional liars; people who might say anything for some hidden purpose. Sometimes we can detect the lies and sometimes they do great harm or damage before it catches up with them.

For some of us, as we go through life and gain experience, knowledge, or wisdom, the built-in lie and bullshit detector develops and fine-tunes itself. This provides defense and protection against the liars who use deception of all types, and for all reasons. It is a valuable tool or defense, worthy of development in men, women, and children; especially children.

Well, we might be wondering what this has to do with building the future; here is a chance to try on your built-in lie detector. On one hand it has nothing to do with the actual construction of the future and, on the other, it has everything to do with the future. A future built on right ways of living, good, and truth should make better and more enjoyable living conditions, kind of like a little piece of heaven. A future built on lies, deception, evil, etc., will probably produce a little piece of hell. It is all just about that simple.

Just so we do not forget, every once in a while, try to picture yourself on this big rock called Earth. It hangs mysteriously suspended in space; it goes around and around at about 1,000 miles per hour; it flies around the Sun at about sixty-six-thousand miles per hour; and science tells us it moves around the galaxy, something like once every two hundred and fifty million years.

It is all very, very amazing and miraculous! The words in the Bible tell us God was pleased with this place he created. If you want or need to, you can find understanding, faith, and strength with the unknown supernatural power of God.